Enzyme technology

M. F. CHAPLIN

Reader, Department of Biotechnology
South Bank Polytechnic, London

AND

C. BUCKE

Professor of Biotechnology,
Polytechnic of Central London, London

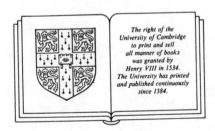

The right of the
University of Cambridge
to print and sell
all manner of books
was granted by
Henry VIII in 1534.
The University has printed
and published continuously
since 1584.

CAMBRIDGE UNIVERSITY PRESS

CAMBRIDGE

NEW YORK   PORT CHESTER   MELBOURNE   SYDNEY

Published by the Press Syndicate of the University of Cambridge
The Pitt Building, Trumpington Street, Cambridge CB2 1RP
40 West 20th Street, New York, NY 10011, USA
10 Stamford Road, Oakleigh, Melbourne 3166, Australia

First published 1990

Printed in Great Britain at the University Press, Cambridge

British Library cataloguing in publication data
Chaplin, M. F.
Enzyme technology.
I. Title
661'.8

Library of Congress cataloguing in publication data
Chaplin, M. F. (Martin F.)
Enzyme technology/M. F. Chaplin and C. Bucke.
    p.   cm.
Bibliography: p.
Includes index.
ISBN 0 521 34429 8. – ISBN 0 521 34884 6 (pbk)
1. Enzymes – Biotechnology.
2. Immobilized enzymes – Biotechnology.
I. Bucke, C.   II. Title.
TP248.65.E59C48 1990
660'.634 – dc20   89-7372 CIP

ISBN 0 521 34429 8 hardback
ISBN 0 521 34884 6 paperback

CE

To my parents June and Frank,
to my wife Philippa
and to my children Timothy, Helen and Katy.

M.F.C.

To my parents.

C.B.

Contents

x   *Contents*

Preface

Enzyme technology has made a major impact on society and is the central theme around which most biotechnological science revolves. In this volume, we have attempted to put it in its rightful place, to show why it is so important, and to show how it is being used at the moment and how it may develop in the future.

This book is intended to be a text for budding and established biotechnologists. It deals with the enzyme technology necessary for diploma, undergraduate and postgraduate students of biotechnology, applied and industrial biology, biochemical engineering, food science and allied subjects. We also hope that it will find favour with their teachers, offering them a fresh approach to the subject area. It is designed to take the reader rapidly through the necessary background and towards a full discussion of the uses of enzymes as industrial catalysts. Overall, we have attempted to give a critical view of current theory, the original references being checked wherever possible. During this process we have come across a number of often-quoted misconceptions, which we hope to dispel.

We attempt to give the kinetic background which we feel is necessary for a proper understanding of the activity of industrial enzymes. In this, we hope that we have achieved a balance between rigour and utility. In particular, we feel that students (and teachers) should, as far as possible, understand the derivation of the equations that are used. The emphasis is on the practical significance of enzyme kinetics. For this reason, enzymic denaturation and the kinetics of reversible reactions have both received significant attention. Throughout, we have tried to instil a feel for the economics of the processes and the reasoning behind practical decision-making.

For each chapter we give a brief summary, containing the 'take-home messages', and a recent bibliography that should enable the interested student (and teacher) rapidly to enter the relevant literature.

We are indebted to a number of our colleagues for their useful discussions and suggestions. In particular we thank Drs T.P. Coultate, E.G. Killick, N.Morgan and M.D. Trevan, Mr S.A. Roulston and Ms H. Shukla.

We apologise for any errors we have inadvertently allowed into print and welcome comments and criticisms from readers.

December 1988   M.F.C.
                                                                    C.B.

Symbols

[]	concentration of the material within the brackets (M)[1]
$[]_0$	concentration at zero time
$[\,_0]$	bulk concentration
α	(alpha), relative volume of organic phase
β	(beta), dimensionless substrate concentration ($= [S]/K_m$)
δ	(delta), effective thickness of the unstirred layer surrounding an immobilised enzyme
ϵ	(epsilon), porosity
η	(eta), (1) effectiveness factor; (2) dynamic viscosity (g m^{-1} s^{-1})
θ	(theta), shape factor
Λ	(lambda), electrostatic partition coefficient
μ	(mu), external substrate modulus
μ_H	proton modulus
μ_i	external substrate modulus in the presence of inhibitor
γ	(nu), kinematic viscosity (m^2 s^{-1})
ρ	(rho), (1) the ratio of radial distance to particle radius (r/R); (2) density (kg l^{-1})
Σ	(sigma), sigma factor (m^2)
τ	(tau), (1) tortuosity; (2) average residence time in reactor
T	(tau), ionic strength (M)
ϕ	(psi), (1) substrate modulus for internal diffusion; (2) volumetric throughput (s^{-1})
ω	(omega) angular velocity (s^{-1})
A	(1) average specific activity; (2) area
C_b	concentration in bottom phase (M)
C_t	concentration in top phase (M)
d	particle diameter
D	(1) diffusivity (m^2 s^{-1}); (2) sedimentation distance
D_F	filter cake thickness

[1] Where appropriate, the most commonly encountered units are given (g, gram; h, hour; J, joule; kg, kilogram; l, litre (dm^3); M, mol l^{-1}; m, metre; min, minute; mol, mole; s, second).

D_S	diffusivity of a substrate $(\text{m}^2\,\text{s}^{-1})$
D_P	diffusivity of a product $(\text{m}^2\,\text{s}^{-1})$
DE	dextrose equivalent
DH	degree of hydrolysis
E	enzyme
ES	enzyme–substrate complex
EP	enzyme–product complex
f	(1) fluid velocity $(\text{m}\,\text{s}^{-1})$; (2) frequency of oscillation
f_m	mass flow rate $(\text{kg}\,\text{m}^{-2}\,\text{s}^{-1})$
F	(1) flow rate $(\text{l}\,\text{s}^{-1})$; (2) Faraday $(= 96487$ coulombs molecular equivalent$^{-1})$
g	gravitational acceleration $(= 981\,\text{cm}\,\text{s}^{-2})$
ΔG^*	standard free energy of activation $(\text{kJ}\,\text{mol}^{-1})$
ΔH	change in enthalpy $(\text{kJ}\,\text{mol}^{-1})$
$[\text{H}_\text{o}^+]$	bulk hydrogen ion concentration (M)
i	degree of backmixing
k	(1) rate constants (units vary); (2) proportionality constant
k_{cat}	catalytic rate constant (s^{-1})
k_d	first-order inactivation constant (s^{-1})
k_L	mass transfer coefficient $(\text{m}\,\text{s}^{-1})$
k_L^{HB}	mass transfer coefficient of the conjugate acid HB $(\text{m}\,\text{s}^{-1})$
k_L^{H}	mass transfer coefficient of the hydrogen ion $(\text{m}\,\text{s}^{-1})$
k_L^{S}	mass transfer coefficient of the substrate $(\text{m}\,\text{s}^{-1})$
k_L^{P}	mass transfer coefficient of the product $(\text{m}\,\text{s}^{-1})$
K	kinetic constant in reversible reactions (M)
K_a	acid dissociation constant
K_{biphasic}	apparent equilibrium constant of biphasic system
K_{eq}	equilibrium constant
K_m	Michaelis constant (M)
K_m^{app}	apparent Michaelis constant (M)
K_{org}	equilibrium constant in an organic solvent
K_S	substrate inhibition constant (M)
K_P	product inhibition constant (M)
K_W	equilibrium constant in water
L	characteristic length of a system
Lf	Le Goff number
$\text{Log}\,P$	solvent polarity
Re	Reynolds number
pI	Isoelectric point
$\text{p}K_\text{a}$	$-\log_{10}$ (acid dissociation constant)

P	product
P	(1) protein content; (2) partition coefficient; (3) pressure through a filter
P_∞	product present at equilibrium
P_0	product, at time zero
P_m	maximum protein releasable
P_r	protein released
P_t	protein remaining at time t
P_t	total product present
P_P	partition coefficient of the product
P_S	partition coefficient of the substrate
Q_{10}	increase in the rate of a reaction per 10 deg. C rise in temperature
r	(1) radial distance within a porous biocatalyst; (2) radius of rotation
R	(1) radius of a porous biocatalyst (m); (2) gas constant ($= 8.314$ J K^{-1} mol^{-1}); (3) resistance of thermistor
S	substrate
S	solubility (g l^{-1})
$S_{1/2}$	substrate present when an enzyme-catalysed reaction is at half the maximal rate
S^*	substrate that can react in a reversible reaction
S_0	substrate, at zero time
S_r	substrate present at radius r within an enzyme particle
S_R	substrate present at the surface of an enzyme particle of radius R
S_t	substrate present at time t
S_t	total substrate present
S_∞	substrate present at equilibrium
t	time
$t_{1/2}$	enzyme half life
T	absolute temperature (K)
U	enzyme activity unit
v	(1) rate of reaction (M s^{-1}); (2) rate of sedimentation (kg cm^{-1} s^{-1})
v_{free}	rate of reaction of a non-immobilised enzyme (mol s^{-1})
V	(1) kinetic constant in reversible reactions (mol s^{-1}); (2) volume of centrifuge (l); (3) volume of reactor
V^*	kinetic constant, maximum (hypothetical) rate of reaction with respect to the pH (mol s^{-1})
V^f	rate of forward reaction (mol s^{-1})
V_b	volume in bottom phase
V_{max}	maximum rate of reaction (mol s^{-1})
V_{max}^{app}	apparent maximum rate of reaction (mol s^{-1})

V_{org}	volume of the organic phase
V^r	rate of backward reaction (mol s^{-1})
V_t	volume in top phase
V_w	volume of aqueous phase
Vol$_S$	volume of a batch reactor
X	fractional conversion
X_o	initial fractional conversion
X_t	fractional conversion at time t

1   Fundamentals of enzyme kinetics

Why enzymes?

Catalysts increase the rate of otherwise slow or imperceptible reactions, without undergoing any net change in their structure. The early development of the concept of catalysis in the nineteenth century went hand in hand with the discovery of powerful catalysts from biological sources. These were called enzymes and were later found to be proteins. They mediate all synthetic and degradative reactions carried out by living organisms. They are very efficient catalysts, often far superior to conventional chemical catalysts, for which reason they are being employed increasingly in today's high-technological society, as a highly significant part of the biotechnological expansion. Their utilisation has created a billion dollar business, including a wide diversity of industrial processes, consumer products, and the burgeoning field of biosensors. Further applications are being discovered constantly.

Enzymes have a number of distinct advantages over conventional chemical catalysts. Foremost amongst these are their specificity and selectivity not only for particular reactions but also in their discrimination between similar parts of molecules (*regiospecificity*) or between optical isomers (*stereospecificity*). They catalyse only the reactions of very narrow ranges of reactants (*substrates*), which may consist of a small number of closely related classses of compounds (e.g. trypsin catalyses the hydrolysis of some peptides and esters in addition to most proteins), a single class of compounds (e.g. hexokinase catalyses the transfer of a phosphate group from ATP to several hexoses), or a single compound (e.g. glucose oxidase oxidises only D-glucose amongst the naturally occurring sugars). This means that the chosen reaction can be catalysed to the exclusion of side-reactions, eliminating undesirable by-products. Thus, higher productivities may be achieved, reducing material costs. As a bonus, the product is generated in an uncontaminated state, so reducing purification costs and the downstream environmental burden. Often a smaller number of steps may be required to produce the desired end-product. In addition, certain stereospecific reactions (e.g. the conversion of glucose into fructose) cannot be achieved by classical chemical methods without a large expenditure of time and effort. Enzymes work under generally mild processing conditions of temperature, pressure and pH. This decreases the energy requirements, reduces the capital costs resulting from corrosion-resistant process equipment and further reduces unwanted side-reactions. The high reaction velocities and straightforward catalytic regulation achieved in enzyme-catalysed reactions allow an increase in productivity, with reduced manufacturing costs due to wages and overheads.

There are some disadvantages in the use of enzymes which cannot be ignored but which are currently being addressed and overcome. In particular, the high cost of enzyme isolation and purification still discourages their use, especially in areas which currently have an established alternative procedure. The generally unstable nature of enzymes, when removed from their natural environment, is also a major drawback to their more extensive use.

Enzyme nomenclature

All enzymes contain a protein backbone. In some enzymes this is the only component in the structure. However there are additional non-protein moieties usually present which may or may not participate in the catalytic activity of the enzyme. Covalently attached carbohydrate groups are commonly encountered structural features which often have no direct bearing on the catalytic activity, although they may well affect an enzyme's stability and solubility. Other factors often found are metal ions (*cofactors*) and low molecular weight organic molecules (*coenzymes*). These may be loosely or tightly bound by non-covalent or covalent forces. They are often important constituents contributing to both the activity and stability of the enzymes. This requirement for cofactors and coenzymes must be recognised if the enzymes are to be used efficiently and is particularly relevant in continuous processes where there may be a tendency for them to become separated from an enzyme's protein moiety.

Enzymes are classified according to the report of the Nomenclature Committee appointed by the International Union of Biochemistry (1984). This enzyme commission assigned each enzyme a recommended name and a four-part distinguishing number. It should be appreciated that some alternative names remain in such common usage that they will be used, where appropriate, in this text. The Enzyme Commission (EC) numbers divide enzymes into six main groups according to the type of reaction catalysed:

(1) *Oxidoreductases*, which involve redox reactions in which hydrogen or oxygen atoms or electrons are transferred between molecules. This extensive class includes the dehydrogenases (hydride transfer), oxidases (electron transfer to molecular oxygen), oxygenases (oxygen transfer from molecular oxygen) and peroxidases (electron transfer to peroxide). For example: glucose oxidase (EC 1.1.3.4, systematic name, β-D-glucose : oxygen 1-oxidoreductase):

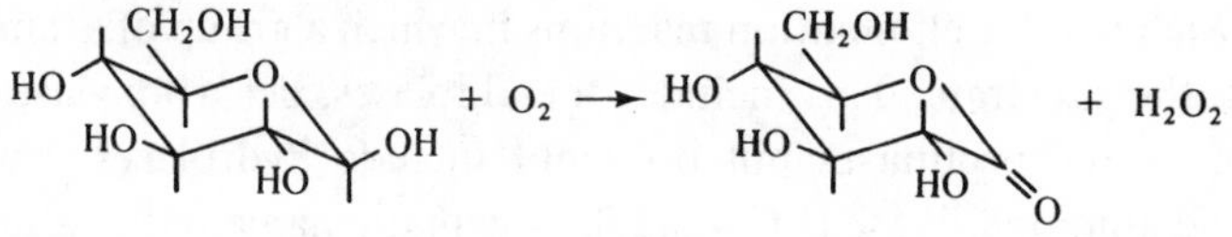

$$\beta\text{-D-glucose} + \text{oxygen} \longrightarrow \text{D-glucono-1,5-lactone} + \text{hydrogen peroxide}$$

[1.1]

(2) *Transferases*, which catalyse the transfer of an atom or group of atoms (e.g. acyl-, alkyl- and glycosyl-groups), between two molecules, but excluding such transfers as are classified in the other groups (e.g. oxidoreductases and hydrolases). For example: aspartate aminotransferase (EC 2.6.1.1, systematic name, L-aspartate : 2-oxoglutarate aminotransferase; also called glutamic-oxaloacetic transaminase or simply GOTase):

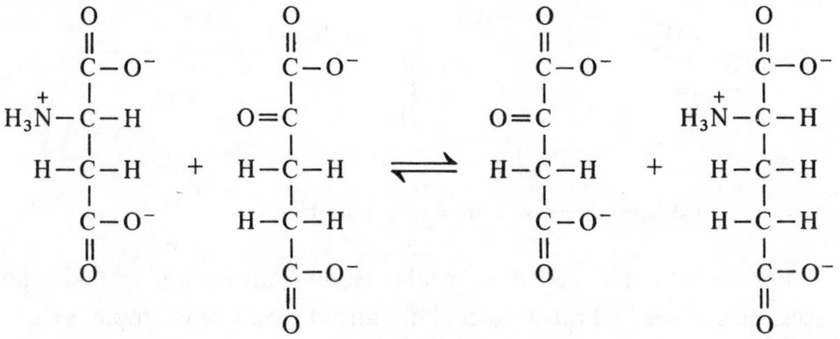

[1.2]

$$\text{L-aspartate} \;+\; \text{2-oxoglutarate} \rightleftharpoons \text{oxaloacetate} \;+\; \text{L-glutamate}$$

(3) *Hydrolases*, which involve hydrolytic reactions and their reversal. This is presently the most commonly encountered class of enzymes within the field of enzyme technology and includes the esterases, glycosidases, lipases and proteases. For example: chymosin (EC 3.4.23.4, no systematic name declared; also called rennin):

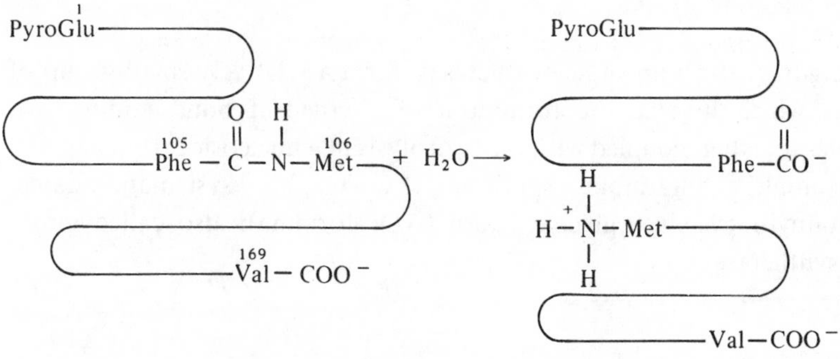

[1.3]

$$\kappa\text{-casein} + \text{water} \longrightarrow para\text{-}\kappa\text{-casein} + \text{caseino macropeptide}$$

(4) *Lyases*, which involve elimination reactions in which a group of atoms is removed from the substrate. This includes the aldolases, decarboxylases, dehydratases and some pectinases but does not include hydrolases. For example: histidine ammonia-lyase (EC 4.3.1.3, systematic name, L-histidine ammonia-lyase; also called histidase);

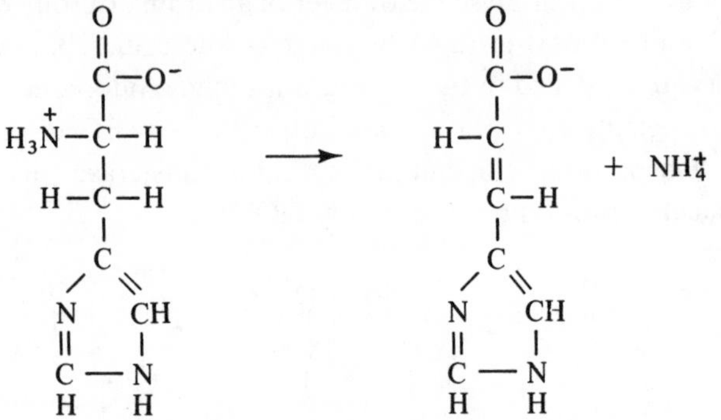

$$
\begin{array}{ccc}
\underset{\substack{| \\ H_3\overset{+}{N}-C-H \\ | \\ H-C-H \\ | \\ \text{ring}}}{\overset{O}{\underset{\|}{C}}-O^-} & \longrightarrow & \underset{\substack{| \\ H-C \\ \| \\ C-H \\ | \\ \text{ring}}}{\overset{O}{\underset{\|}{C}}-O^-} + NH_4^+
\end{array}
\qquad [1.4]
$$

L-histidine ⟶ urocanate + ammonia

(5) *Isomerases*, which catalyse molecular isomerisations and include the epimerases, racemases and intramolecular transferases. For example: xylose isomerase (EC 5.3.1.5, systematic name, D-xylose ketol-isomerase; commonly called glucose isomerase):

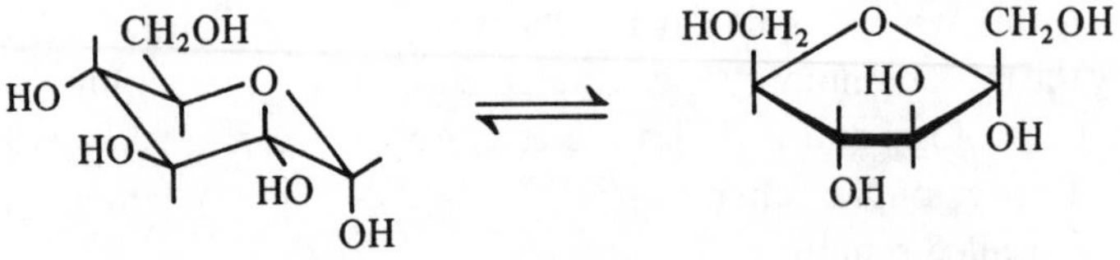

[1.5]

α-D-glucopyranose ⇌ α-D-fructofuranose

(6) *Ligases*, also known as synthetases, form a relatively small group of enzymes which involve the formation of a covalent bond joining two molecules together, coupled with the hydrolysis of a nucleoside triphosphate. For example: glutathione synthase (EC 6.3.2.3, systematic name, γ-L-glutamyl-L-cysteine : glycine ligase (ADP-forming); also called glutathione synthetase):

$$\text{ATP} + \begin{array}{c} \text{O} \quad \text{CH}_2\text{SH} \\ \| \quad | \\ \text{C}-\text{NHCHC}-\text{O}^- \\ | \qquad \| \\ \text{H}-\text{C}-\text{H} \quad \text{O} \\ | \\ \text{H}-\text{C}-\text{H} \\ | \\ \text{H}-\text{C}-\text{NH}_3^+ \\ | \\ \text{C}-\text{O}^- \\ \| \\ \text{O} \end{array} + \begin{array}{c} \text{O} \\ \| \\ \text{C}-\text{O}^- \\ | \\ \overset{+}{\text{NH}_3}-\text{C}-\text{H} \\ | \\ \text{H} \end{array} \longrightarrow \text{ADP} + \text{H}_2\text{PO}_4^- + \begin{array}{c} \text{O} \quad \text{CH}_2\text{SH} \quad \text{O} \\ \| \quad | \qquad \| \\ \text{C}-\text{NHCHCONHCH}_2\text{C}-\text{O}^- \\ | \\ \text{H}-\text{C}-\text{H} \\ | \\ \text{H}-\text{C}-\text{H} \\ | \\ \text{H}-\text{C}-\text{NH}_3^+ \\ | \\ \text{C}-\text{O}^- \\ \| \\ \text{O} \end{array}$$

[1.6]

ATP + γ-L-glutamyl-L-cysteine + glycine ⟶ ADP + phosphate + glutathione

Enzyme units

The amount of enzyme present or used in a process is difficult to determine in absolute terms (e.g. grams), as its purity is often low and a proportion may be in an inactive, or partially active, state. More relevant parameters are the activity of the enzyme preparation and the activities of any contaminating enzymes. These activities are usually measured in terms of the *activity unit* (U), which is defined as the amount which will catalyse the transformation of 1 μmole of the substrate per minute under standard conditions. Typically, this represents 10^{-6}–10^{-11} kg for pure enzymes and 10^{-4}–10^{-7} kg for industrial enzyme preparations. Another unit of enzyme activity has been recommended. This is the *katal* (kat), which is defined as the amount which will catalyse the transformation of one mole of substrate per second (1 kat = 60000000 U). It is an impracticable unit and has not yet received widespread acceptance. Sometimes non-standard activity units are used, such as Soxhlet, Anson and Kilo Novo units, which are based on physical changes such as lowering viscosity and are supposedly better understood by industry. Rightfully, such units are gradually falling into disuse. The activity is a measure of enzyme content that is clearly of major interest when the enzyme is to be used in a process. For this reason, enzymes are usually marketed in terms of activity rather than weight. The specific activity (e.g. U kg^{-1}) is a parameter of interest, having some utility as an index of purity but of lesser importance than activity. There is a major problem with these definitions of activity: the rather vague notion of 'standard conditions'. These are meant to refer to optimal conditions, especially with regard to pH, ionic strength, temperature, substrate concentration and the presence and concentration of cofactors and coenzymes. However, these so-termed optimal conditions vary both between laboratories and between suppliers. They also depend on the particular application in which the enzyme is to be used. Additionally, preparations of the same notional specific activity may

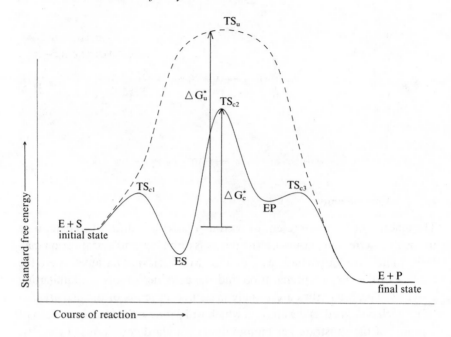

Figure 1.1. Diagram showing the free-energy profile of the course of an enzyme-catalysed reaction involving the formation of enzyme–substrate (ES) and enzyme–product (EP) complexes (i.e. $E + S \rightleftharpoons ES \rightleftharpoons EP \rightleftharpoons P + E$). The catalysed reaction pathway goes through the transition states TS_{c1}, TS_{c2} and TS_{c3}, with standard free energy of activation ΔG_c^*, whereas the uncatalysed reaction goes through the transition state TS_u with standard free energy of activation ΔG_u^*. In this example the rate-limiting step would be the conversion of ES into EP. Reactions involving several substrates and products, or more intermediates, are even more complicated. The Michaelis–Menten reaction scheme [1.7] would give a similar profile but without the EP-complex free energy trough. The schematic profile for the uncatalysed reaction is shown as the dashed line. It should be noted that the catalytic effect only concerns the lowering of the standard free energy of activation from ΔG_u^* to ΔG_c^* and has no effect on the overall free energy change (i.e. the difference between the initial and final states) or the related equilibrium constant.

differ with respect to stability and be capable of very different total catalytic *productivity* (this is the total substrate converted to product during the lifetime of the catalyst, under specified conditions). Conditions for maximum initial activity are not necessarily those for maximum stability. Great care has to be taken over the consideration of these factors when the most efficient catalyst for a particular purpose is to be chosen.

The mechanism of enzyme catalysis

In order for a reaction to occur, reactant molecules must contain sufficient energy to cross a potential energy barrier, the *activation energy*. All molecules possess varying amounts of energy depending, for example, on their recent collision history, but generally only a few have sufficient energy for reaction. The lower the potential energy barrier to reaction, the more reactants have sufficient energy and, hence, the faster the reaction will occur. All catalysts, including enzymes, function by forming a transition state, with the reactants, of lower free energy than would be found in the uncatalysed reaction (Figure 1.1). Even quite modest reductions in this potential energy barrier may produce large increases in the rate of reaction (e.g. the activation energy for the uncatalysed breakdown of hydrogen peroxide to oxygen and water is 76 kJ mol^{-1} whereas, in the presence of the enzyme catalase, this is reduced to 30 kJ mol^{-1} and the rate of reaction is increased by a factor of 10^8, sufficient to convert a reaction time measured in years into one measured in seconds).

There is a number of mechanisms by which this activation energy decrease may be achieved. The most important of these involves the enzyme initially binding the substrate(s), in the correct orientation to react, close to the catalytic groups on the active enzyme complex. In this way the binding energy is used partly to reduce the contribution, towards the total activation energy, of the considerable activation entropy due to the loss of the reactants' (and catalytic groups') translational and rotational entropy. Other contributing factors are the introduction of strain into the reactants (allowing more binding energy to be available for the transition state), provision of an alternative reactive pathway and the desolvation of reacting and catalysing ionic groups.

The energies available to enzymes for binding their substrates are determined primarily by the complementarity of structures (i.e. a good three-dimensional fit plus optimal non-covalent ionic and/or hydrogen-bonding forces). The specificity depends upon minimal steric repulsion, the absence of unsolvated or unpaired charges, and the presence of sufficient hydrogen bonds. These binding energies may be quite large: for example, antibody–antigen dissociation constants are characteristically near 10^{-8} M (free energy of binding is 46 kJ mol^{-1}); ATP binds to myosin with a dissociation constant of 10^{-13} M (free energy of binding is 75 kJ mol^{-1}); and biotin binds to avidin, a protein found in egg white, with a dissociation constant of 10^{-15} M (free energy of binding is 86 kJ mol^{-1}). However, enzymes do not use this potential binding energy simply to bind the substrate(s) and form stable long-lasting complexes. If this were to be the case, the formation of the transition state between the enzyme–substrate complex (ES) and the enzyme–product complex (EP) would involve an extremely large free energy change due to the breaking of these strong binding forces, and the rate of formation of products would be very slow. They must use this binding energy for reducing the free energy of the transition state. This is generally achieved by increasing the binding to the transition state rather than to the reactants and, in the process, introducing an energetic strain into the system and allowing more favourable interactions between the enzyme's catalytic groups and the reactants.

Simple kinetics of enzyme action

It is established that enzymes form a bound complex with their reactants (i.e. *substrates*) during the course of their catalysis and prior to the release of products. This can be simply illustrated, using a mechanism based on that of Michaelis and Menten for a one-substrate reaction, by the reaction sequence:

enzyme + substrate $\rightleftharpoons$ (enzyme–substrate complex) $\longrightarrow$ enzyme + product

$$E + S \underset{k_{-1}}{\overset{k_{+1}}{\rightleftharpoons}} ES \overset{k_{+2}}{\longrightarrow} P \qquad [1.7]$$

where k_{+1}, k_{-1} and k_{+2} are the respective rate constants, typically having values of $10^5–10^8 \, M^{-1} \, s^{-1}$, $1 - 10^4 \, s^{-1}$ and $1–10^5 \, s^{-1}$, respectively, the sign of the subscripts indicating the direction in which the rate constant is acting. For the sake of simplicity, the reverse reaction (the conversion of product to substrate) is not included in this scheme. This is allowable (1) at the beginning of the reaction when there is no, or little, product present, or (2) when the reaction is effectively irreversible. Reversible reactions are dealt with in more detail later in this chapter.

The rate of reaction (v) is the rate at which the product is formed:

$$v = \frac{d[P]}{dt} = k_{+2}[ES] \qquad (1.1)$$

where [] indicates the molar concentration of the material enclosed (i.e. [ES] is the concentration of the enzyme–substrate complex). The rate of change of the concentration of ES equals the rate of its formation minus the rate of its breakdown, forwards to give product or backwards to regenerate substrate:

therefore:

$$\frac{d[ES]}{dt} = k_{+1}[E][S] - (k_{-1} + k_{+2})[ES] \qquad (1.2)$$

During the course of the reaction, the total enzyme at the beginning of the reaction ($[E]_0$, at zero time) is present either as the free enzyme ([E]) or the ES complex ([ES]):

i.e.: $\quad [E]_0 = [E] + [ES]$ (1.3)

therefore:

$$d[ES]/dt = k_{+1}([E]_0 - [ES])[S] - (k_{-1} + k_{+2})[ES]$$ (1.4)

Gathering terms together, this gives:

$$\frac{d[ES]/dt}{k_{+1}[S] + k_{-1} + k_{+2}} + [ES] = \frac{k_{+1}[S][E]_0}{k_{+1}[S] + k_{-1} + k_{+2}}$$ (1.5)

The differential equation (1.5) is difficult to handle, but may be greatly simplified if it can be assumed that the left-hand side is equal to [ES] alone. This assumption is valid under the sufficient but unnecessarily restrictive steady-state approximation that the rate of formation of ES equals its rate of disappearance by product formation and reversion to substrate (i.e. $d[ES]/dt$ is zero). It is additionally valid when the condition:

$$\frac{d[ES]/dt}{k_{+1}[S] + k_{-1} + k_{+2}} \ll [ES]$$ (1.6)

is valid. This occurs during a substantial part of the reaction time-course over a wide range of kinetic rate constants and substrate concentrations and at low to moderate enzyme concentrations. The variation in [ES], $d[ES]/dt$, [S] and [P] with the time-course of the reaction is shown in Figure 1.2, where it may be seen that the simplified equation is valid throughout most of the reaction.

The Michaelis–Menten equation (below) is simply derived from equations (1.1) and (1.5), by substituting K_m for $(k_{-1} + k_{+2})/k_{+1}$. K_m is known as the *Michaelis constant* with a value typically in the range 10^{-1}–10^{-5} M. When $k_{+2} \ll k_{-1}$, K_m equals the dissociation constant (k_{-1}/k_{+1}) of ES:

$$v = k_{+2}[ES] = \frac{k_{+2}[S][E]_0}{[S] + K_m}$$ (1.7)

or, more simply:

$$v = \frac{V_{max}[S]}{[S] + K_m}$$ (1.8)

where V_{max} is the maximum rate of reaction, which occurs when the enzyme is completely saturated with substrate (i.e. when [S] is very much greater than K_m, V_{max} equals $k_{+2}[E]_0$, as the maximum value [ES] can have is $[E]_0$ when $[E]_0$ is less than $[S]_0$). Equation (1.8) may be rearranged to show the dependence of the rate of reaction on the ratio of $[S] : K_m$:

$$v = \frac{V_{max}}{1 + K_m/[S]}$$ (1.9)

and the rectangular hyperbolic nature of the relationship, having asymptotes at $v = V_{max}$ and $[S] = -K_m$:

$$(V_{max} - v)(K_m + [S]) = V_{max}K_m$$ (1.10)

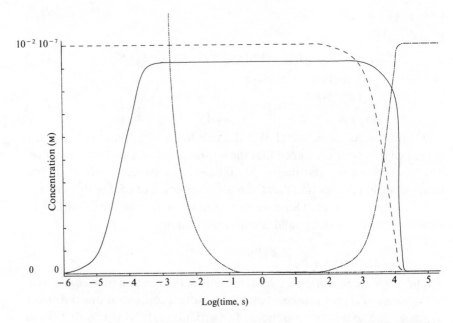

Figure 1.2. Computer simulation of the progress curves of [ES], d[ES]/dt, [S] and [P] for a reaction obeying simple Michaelis–Menten kinetics with $k_{+1} = 10^6$ M^{-1} s^{-1}, $k_{-1} = 1000$ s^{-1}, $k_{+2} = 10$ s^{-1}, [E]$_0 = 10^{-7}$ M and [S]$_0 = 0.01$ M. —— [ES] (0–10^{-7} M scale), ···—···—··· d[ES]/dt (0–10^{-7} M scale), ––––– [S] (0–10^{-2} M scale), —·—·— [P] (0–10^{-2} M scale). The simulation shows three distinct phases to the reaction time-course. There is an initial transient phase, which lasts for about a millisecond, followed by a longer steady-state phase of about 30 min, when [ES] stays constant but only a small proportion of the substrate reacts. This is followed by the final phase, taking about 6 h, during which the substrate is completely converted to product.
(d[ES]/dt)/(k_{+1}[S] + k_{-1} + k_2) is much less than [ES] during both of the latter two phases.

The substrate concentration in these equations is the actual concentration at the time and, in a closed system, will only be approximately equal to the initial substrate concentration ([S]$_0$) during the early phase of the reaction. Hence, it is usual to use these equations to relate the initial rate of reaction to the initial, and easily predetermined, substrate concentration (Figure 1.3). This also avoids any problem that may occur through product inhibition or reaction reversibility (see later).

It has been established that few enzymes follow the Michaelis–Menten equation over a wide range of experimental conditions. However, it remains by far the most generally applicable equation for describing enzymic reactions. Indeed it can be applied realistically to a number of reactions

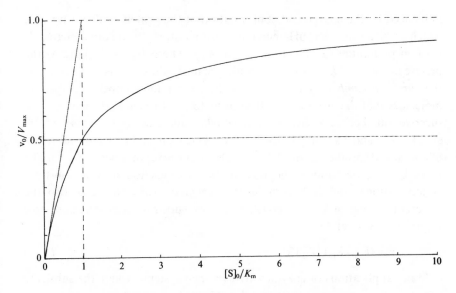

Figure 1.3. A normalised plot of the initial rate (v_0) against initial substrate concentration ($[S]_0$) for a reaction obeying Michaelis–Menten kinetics (equation (1.8)). The plot has been normalised in order to make it more generally applicable by plotting the relative initial rate of reaction (v_0/V_{max}) against the initial substrate concentration relative to the Michaelis constant ($[S]_0/K_m$, more commonly referred to as β, the dimensionless substrate concentration). The curve is a rectangular hyperbola with asymptotes at $v_0 = V_{max}$ and $[S]_0 = -K_m$. The tangent to the curve at the origin goes through the point ($v_0 = V_{max}$), ($[S]_0 = K_m$). The ratio V_{max}/K_m is an important kinetic parameter which describes the relative specificity of a fixed amount of the enzyme for its substrate (more precisely defined in terms of k_{cat}/K_m). The substrate concentration, which gives a rate of half the maximum reaction velocity, is equal to the K_m.

which have a far more complex mechanism than the one described here. In these cases K_m remains an important quantity, characteristic of the enzyme and substrate, corresponding to the substrate concentration needed for half the enzyme molecules to bind to the substrate (and, therefore, causing the reaction to proceed at half its maximum rate); the precise kinetic meaning derived earlier, however, may not hold and may be misleading. In these cases the K_m is likely to equal a much more complex relationship between the many rate constants involved in the reaction scheme. It remains independent of the enzyme and substrate concentrations and indicates the extent of binding between the enzyme and its substrate for a given substrate concentration, a lower K_m indicating a greater extent of binding. V_{max} clearly depends on the enzyme concentration and for some, but not all, enzymes may be largely independent of the specific substrate used: K_m and V_{max} may both be influenced by the charge and conformation of the protein and substrate(s), which are determined by pH, temperature, ionic strength and other factors. It is often preferable to substitute k_{cat} for k_{+2}, where $V_{max} = k_{cat}[E]_0$, as the precise meaning of k_{+2}, above, may also be misleading. k_{cat} is also known as the *turnover number* as it represents the maximum number of substrate molecules that the enzyme can 'turn over' to product in a set time (e.g. the turnover numbers of α-amylase, glucoamylase and glucose isomerase are 500 s^{-1}, 160 s^{-1} and 3 s^{-1}, respectively; an enzyme with a relative molecule mass of 60 000 and specific activity 1 U mg^{-1} has a turnover number of 1 s^{-1}). The ratio k_{cat}/K_m determines the relative rate of reaction at low substrate concentrations, and is known as the *specificity constant*. It is also the apparent second-order rate constant at low substrate concentrations (see Figure 1.3), where:

$$v = (k_{cat}/K_m) [E]_0[S] \tag{1.11}$$

Many applications of enzymes involve open systems, where the substrate concentration remains constant, due to replenishment, throughout the time-course of the reaction. This is, of course, the situation that often prevails *in vivo*. Under these circumstances, the Michaelis–Menten equation is obeyed over an even wider range of enzyme concentrations than is allowed in closed systems, and is commonly used to model immobilised enzyme kinetic systems (see Chapter 3).

Enzymes have evolved by maximising k_{cat}/K_m (i.e. the specificity constant for the substrate), whilst keeping K_m approximately identical with the naturally encountered substrate concentration. This allows the enzyme to operate efficiently and yet exercise some control over the rate of reaction.

The specificity constant is limited by the rate at which the reactants encounter one another under the influence of diffusion. For a single-substrate reaction, the rate of encounter between the substrate and enzyme is about 10^8–10^9 M^{-1} s^{-1}. The specificity constant of some enzymes approaches this value, although the range of determined values is very broad (e.g. k_{cat}/k_m is 4×10^7 M^{-1} s^{-1} for catalase, but 25 M^{-1} s^{-1} for glucose isomerase, and for other enzymes it varies from less than 1 M^{-1} s^{-1} to greater than 10^8 M^{-1} s^{-1}).

Effect of pH and ionic strength on enzyme catalysis

Enzymes are amphoteric molecules containing a large number of acidic and basic groups, situated mainly on the surface. The charges on these groups will vary, according to their acid dissociation constants, with the pH of their environment (Table 1.1). This will affect the total net charge of the enzyme and the distribution of charge on its exterior surface, in addition to the

Table 1.1 pK_a values[a] and heats of ionisations[b] of the ionising groups commonly found in enzymes

Group	Usual pK_a range	Approximate charge at pH 7	Heats of ionisation (kJ mol^{-1})
Carboxyl (C-terminal, glutamic acid, aspartic acid)	3– 6	– 1.0	± 5
Ammonio (N-terminal)	7– 9	+ 1.0	+ 45
(lysine)	9–11	+ 1.0	+ 45
Imidazolyl (histidine)	5– 8	+ 0.5	+ 30
Guanidyl (arginine)	11–13	+ 1.0	+ 50
Phenolic (tyrosine)	9–12	0.0	+ 25
Thiol (cysteine)	8–11	0.0	+ 25

[a] The pK_a (defined as $- \log (K_a)$) is the pH at which half the groups are ionised. Note the similarity between the K_a of an acid and the K_m of an enzyme, which is the substrate concentration at which half the enzyme molecules have bound substrate.

[b] By convention, the heat (enthalpy) of ionisation is positive when heat is withdrawn from the surrounding solution (i.e. the reaction is endothermic) by the dissociation of the hydrogen ions.

reactivity of the catalytically active groups. These effects are especially important in the neighbourhood of the active site. Taken together, the changes in charges with pH affect the activity, structural stability and solubility of the enzyme.

There will be a pH, characteristic of each enzyme, at which the net charge on the molecule is zero. This is called the *isoelectric point* (pI), at which the enzyme generally has minimum solubility in aqueous solutions. In a similar manner to the effect on enzymes, the charge and charge distribution on the substrate(s), product(s) and coenzymes (where applicable) will also be affected by pH changes. Increasing hydrogen ion concentration will, additionally, increase the successful competition of hydrogen ions for any metal cationic binding sites on the enzyme, reducing the bound metal cation concentration. Decreasing hydrogen ion concentration, on the other hand, leads to increasing concentration of hydroxyl ions which compete against the enzymes' ligands for divalent and trivalent cations: this causes conversion of the metal ions to hydroxides and, at high hydroxyl concentrations, their complete removal from the enzyme. The temperature also has a marked effect on ionisations, the extent of which depends on the heats of ionisation of the particular groups concerned (Table 1.1). The relationship between the change in the pK_a and the change in temperature is given by a derivative of the Gibbs–Helmholtz equation:

$$\frac{d(pK_a)}{dT} = \frac{-\Delta H}{2.303RT^2} \qquad (1.12)$$

where T is the absolute temperature (in K), R is the gas law constant (8.314 J K^{-1} mol^{-1}), ΔH is the heat of ionisation and the numeric constant (2.303) is the natural logarithm of 10, as pK_a values are based on logarithms to base 10. This variation is sufficient to shift the pI of enzymes by up to one unit towards lower pH on increasing the temperature by 50 deg.C.

These charge variations, plus any consequent structural alterations, may be reflected in changes in the binding of the substrate, the catalytic efficiency and the amount of active enzyme. Both V_{max} and K_m will be affected due to the resultant modifications to the kinetic rate constants k_{+1}, k_{-1} and k_{cat} (k_{+2} in the Michaelis–Menten mechanism), and the variation in the concentration of active enzyme. The effect of pH on the V_{max} of an enzyme-catalysed reaction may be explained using the, generally true, assumption that only one charged form of the enzyme is optimally catalytic and therefore the maximum concentration of the enzyme–substrate intermediate cannot be greater than the concentration of this species. In simple terms, assume EH^- is the only active form of the enzyme:

$$S + EH^- \underset{k_{-1}}{\overset{k_{+1}}{\rightleftharpoons}} EH^-S \xrightarrow{k_{+2}} P \qquad [1.8]$$

The concentration of EH^- is determined by the two dissociations:

$$EH_2 \overset{K_{a1}}{\rightleftharpoons} EH^- + H^+ \qquad [1.9]$$

$$EH^- \overset{K_{a2}}{\rightleftharpoons} E^{2-} + H^+ \qquad [1.10]$$

with:

$$K_{a1} = \frac{[EH^-]_0[H^+]}{[EH_2]_0} \qquad (1.13)$$

and:

$$K_{a2} = \frac{[E^{2-}]_0[H^+]}{[EH^-]_0} \qquad (1.14)$$

However:

$$[E]_0 = [EH_2]_0 + [EH^-]_0 + [E^{2-}]_0 \qquad (1.15)$$

therefore:

$$[E]_0 = [EH^-]_0\{([H^+]/K_{a1}) + 1 + (K_{a2}/[H^+])\} \qquad (1.16)$$

As the rate of reaction is given by $k_{+2}[EH^-S]$ and this is maximal when $[EH^-S]$ is maximal (i.e. when $[EH^-S] = [EH^-]_0$):

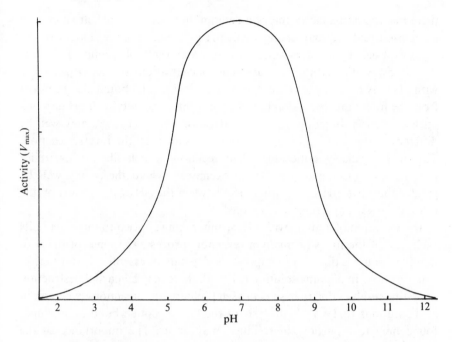

Figure 1.4. A generally applicable diagram of the variation in the rate of an enzyme-catalysed reaction (V_{max}) with the pH of the solution. The centre (optimum pH) and the breadth of this 'bell-shaped' curve depend upon the acid dissociation constants of the relevant groups in the enzyme. It should be noted that some enzymes have pH–activity profiles that show little similarity to this diagram.

$$V_{max} = k_{+2}[EH^-]_0 = \frac{k_{+2}[E]_0}{([H^+]/K_{a1}) + 1 + (K_{a2}/[H^+])} \qquad (1.17)$$

The V_{max} will be greatest when:

$$[H^+]/K_{a1} = K_{a2}/[H^+] \qquad (1.18)$$

therefore:

$$pH_{optimum} = (pK_{a1} + pK_{a2})/2 \qquad (1.19)$$

This derivation has involved a number of simplications of the real situation; it ignores the effect of the ionisation of substrates, products and enzyme–substrate complexes and it assumes that EH^- is a single ionised species, when it may contain a mixture of differently ionised groups but with identical overall charge, although the process of binding substrate will tend to fix the required ionic species. It does, however, produce a variation of maximum rate with pH, which gives the commonly encountered 'bell-shaped' curve (Figure 1.4). Where the actual reaction scheme is more complex, there may be a more complex relationship between V_{max} and pH. In particular there may be a change in the rate-determining step with pH. It should be recognised that K_m may change with pH in a manner independent of the V_{max}, as it usually involves other, or additional, ionisable groups. It is clear that at lower, non-saturating, substrate concentrations the activity changes with pH may or may not reflect the changes in V_{max}. It should also be noted from the foregoing discussion that the variation of activity with pH depends on the reaction direction under consideration. The pH$_{optimum}$ may well be different in the forward direction from that shown by the reverse reaction. This is particularly noticeable when reactions which liberate or utilise protons are considered (e.g. dehydrogenases), where there may well be greater than two pH units' difference between the pH$_{optimum}$ shown by the rates of forward and reverse reactions.

The variation of activity with pH, within a range of two to three pH units each side of the pI, is normally a reversible process. Extremes of pH will, however, cause a time- and temperature-dependent, essentially irreversible, denaturation. In alkaline solution (pH > 8), there may be partial destruction of cystine residues due to base-catalysed β-elimination reactions, whereas, in acid solutions (pH < 4), hydrolysis of the labile peptide bonds, sometimes found next to aspartic acid residues, may occur. The importance of the knowledge concerning the variation of activity with pH cannot be overemphasised. However, a number of other factors may mean that the optimum pH in the V_{max}–pH diagram may not be the pH of choice in a technological process involving enzymes. These include the variation of solubility of substrate(s) and product(s), changes in the position of equilibrium for a reaction, suppression of the ionisation of a product to facilitate its partition and recovery into an organic solvent, and the reduction in susceptibility to oxidation or microbial contamination. The major factor is the effect of pH on enzyme stability. This relationship is further complicated by the variation in the effect of the pH with both the duration of the process and the temperature or temperature–time profile. The important parameter derived from these influences is the productivity of the enzyme (i.e. how much substrate it is capable of converting to product). The variation of productivity with pH may be similar to that of the V_{max}–pH relationship but changes in the substrate stream composition and contact time may also make some contribution. Generally, the variation must be determined under the industrial process conditions.

It is possible to alter the pH–activity profiles of enzymes. Ionisation of the carboxylic acids involves the separation of the released groups of opposite charge. This process is encouraged within solutions of higher polarity and reduced by less polar solutions. Thus, reducing the dielectric constant of an aqueous solution by the addition of a co-solvent of low polarity (e.g. dioxan, ethanol), or by partition on immobilisation (see Chapter 3), increases the pK_a of carboxylic acid groups. Co-solvents are sometimes useful but not generally applicable to enzyme-catalysed reactions, as they may cause a drastic change in an enzyme's productivity due to denaturation (but see Chapter 7). The pK_a of basic groups are not similarly affected, as there is no separation of charges when basic groups ionise. However, protonated basic groups which are stabilised by neighbouring negatively charged groups will be stabilised (i.e. have raised pK_a) by solutions of lower polarity. Changes in the *ionic strength* (*T*) of the solution may also have some effect. The ionic strength is defined as half of the total sum of the concentration (c_i) of every ionic species (*i*) in the solution times the square of its charge (z_i); i.e. $T = 0.5\Sigma(c_i z_i^2)$. For example, the ionic strength of a 0.1 M solution of $CaCl_2$ is 0.3 M, i.e. $0.5 \times [(0.1 \times 2^2) + (0.2 \times 1^2)]$ M. At higher solution ionic strength, charge separation is encouraged with a concomitant lowering of the carboxylic acid pK_a values. These changes, extensive as they may be, have little effect on the overall charge on the enzyme molecule at neutral pH and are, therefore, only likely to exert a small influence on the enzyme's isoelectric point. Chemical derivatisation methods are available for converting surface charges from positive to negative and vice versa. It is found that a single change in charge has little effect on the pH–activity profile, unless it is at the active site. However, if all lysines are converted to carboxylates (e.g. by reaction with succinic anhydride) or if all the carboxylates are converted to amines (e.g. by coupling to ethylene diamine by means of a carbodiimide; see Chapter 3) the profile can be shifted about one pH unit towards higher or lower pH, respectively. The cause of these shifts is primarily the stabilisation or destabilisation of the charges at the active site during the reaction, and the effects are most noticeable at low ionic strength. Some, more powerful, methods for shifting the pH–activity profile are specific to immobilised enzymes and are described in Chapter 3.

The ionic strength of the solution is an important parameter affecting enzyme activity. This is especially noticeable where catalysis depends on the movement of charged molecules relative to each other. Thus, both the binding of charged substrates to enzymes and the movement of charged groups within the catalytic 'active' site will be influenced by the ionic composition of the medium. If the rate of the reaction depends upon the approach of charged moieties, the following approximate relationship may hold:

$$\log (k) = \log (k_0) + z_A z_B T^{1/2} \tag{1.20}$$

where k is the actual rate constant, k_0 is the rate constant at zero ionic strength, z_A and z_B are the electrostatic charges on the reacting species, and T

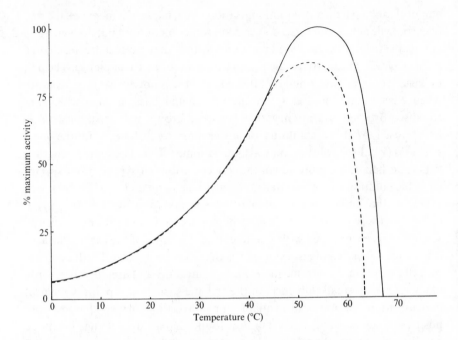

Figure 1.5. Diagram showing the effect of temperature on the activity of an enzyme-catalysed reaction. ——, Short incubation period; ––––, long incubation period. Note that the temperature at which there appears to be maximum activity varies with the incubation time.

is the ionic strength of the solution. If the charges are opposite then there is a decrease in the reaction rate with increasing ionic strength, whereas if the charges are identical, an increase in the reaction rate will occur (e.g. the rate-controlling step in the catalytic mechanism of chymotrypsin involves the approach of two positively charged groups, histidine^{+} [57] and arginine^{+} [145] causing a significant increase in k_{cat} on increasing the ionic strength of the solution). Even if a more complex relationship between the rate constants and the ionic strength holds, it is clearly important to control the ionic strength of solutions in parallel with the control of pH.

Effect of temperature and pressure

Rates of all reactions, including those catalysed by enzymes, rise with increase in temperature in accordance with the Arrhenius equation:

$$k = Ae^{-\Delta G^*/RT} \tag{1.21}$$

where k is the kinetic rate constant for the reaction, A is the Arrhenius constant (also known as the frequency factor), ΔG^* is the standard free

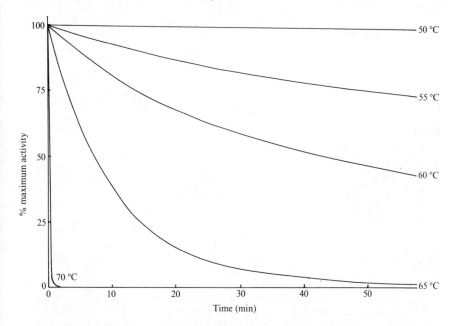

Figure 1.6. Diagram showing the effect of the temperature on the stability of an enzyme-catalysed reaction. The curves show the percentage activity remaining as the incubation period increases. From the top they represent equal increases in the incubation temperature (50, 55, 60, 65 and 70 °C).

energy of activation (kJ mol^{-1}) which depends on entropic and enthalpic factors, R is the gas law constant and T is the absolute temperature. Typical standard free energies of activation (15–70 kJ mol^{-1}) give rise to increases in rate by factors between 1.2 and 2.5 for every 10 deg.C rise in temperature. This factor for the increase in the rate of reaction for every 10 deg.C rise in temperature is commonly denoted by the term Q_{10} (i.e. in this case, Q_{10} is within the range 1.2–2.5). All the rate constants contributing to the catalytic mechanism will vary independently, causing changes in both K_m and V_{max}. It follows that, in an exothermic reaction, the reverse reaction (having a higher activation energy) increases more rapidly with temperature than the forward reaction. This not only alters the equilibrium constant (see equation (1.12)), but also reduces the optimum temperature for maximum conversion as the reaction progresses. The reverse holds for endothermic reactions such as that of glucose isomerase (see reaction [1.5] where the value of the ratio fructose : glucose, at equilibrium, increases from 1.00 at 55 °C to 1.17 at 80 °C.

In general, it would be preferable to use enzymes at high temperatures in order to make use of this increased rate of reaction plus the protection it affords against microbial contamination. Enzymes, however, are proteins

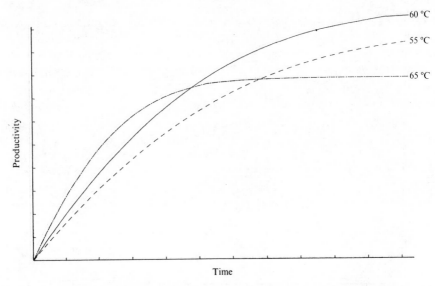

Figure 1.7. Diagram showing the effect of the temperature on the productivity of an enzyme-catalysed reaction. ----- 55 °C; —— 60 °C; —·—·—· 65 °C. The optimum productivity is seen to vary with the process time, which may be determined by other additional factors (e.g. cost of overheads.) It is often difficult to achieve precise control of the temperature of an enzyme-catalysed process and, under these circumstances, it may be seen that it is prudent to err on the low temperature side.

and undergo essentially irreversible *denaturation* (i.e. conformational alteration entailing a loss of biological activity) at temperatures above those to which they are ordinarily exposed in their natural environment. These denaturing reactions have standard free energies of activation of about 200–300 kJ mol^{-1} (Q_{10} in the range 6–36), which means that, above a critical temperature, there is a rapid rate of loss of activity (Figure 1.5). The actual loss of activity is the product of this rate and the duration of incubation (Figure 1.6). It may be due to covalent changes such as the deamination of asparagine residues or non-covalent changes such as the rearrangement of the protein chain. Inactivation by heat denaturation has a profound effect on the enzymes' productivity (Figure 1.7).

The thermal denaturation of an enzyme may be modelled by the following serial deactivation scheme:

$$E \xrightarrow{k_{d1}} E_1 \xrightarrow{k_{d2}} E_2 \qquad\qquad [1.11]$$

where: k_{d1} and k_{d2} are the first-order deactivation rate coefficients; E is the native enzyme, which may, or may not, be an equilibrium mixture of a number of species, distinct in structure or activity; and E_1 and E_2 are enzyme molecules with average specific activity relative to E of A_1 and A_2, respectively. A_1 may be greater or less than unity (i.e. E_1 may have higher or lower activity than E) whereas A_2 is normally very small or zero. This model allows for the rare cases involving free enzyme (e.g. tyrosinase) and the somewhat commoner cases involving immobilised enzyme (see Chapter 3) where there is a small initial activation or period of grace involving negligible discernible loss of activity during short incubation periods but prior to later deactivation. Assuming, at the beginning of the reaction:

$$[E] = [E]_0 \tag{1.22}$$

and:

$$[E_1] = [E_2] = 0 \tag{1.23}$$

At time t:

$$[E] + [E_1] + [E_2] = [E]_0 \tag{1.24}$$

It follows from the reaction scheme [**1.11**]:

$$-d[E]/dt = k_{d1}[E] \tag{1.25}$$

Integrating equation (1.25) using the boundary condition in equation (1.22) gives:

$$[E] = [E]_0 e^{(-k_{d1}t)} \tag{1.26}$$

From the reaction scheme [**1.11**]:

$$-d[E_1]/dt = k_{d2}[E_1] - k_{d1}[E] \tag{1.27}$$

Substituting for [E] from equation (1.26):

$$-d[E_1]/dt = k_{d2}[E_1] - k_{d1}[E]_0 e^{(-k_{d1}t)} \tag{1.28}$$

Integrating equation (1.27) using the boundary condition in equation (1.23) gives:

$$[E_1] = \frac{k_{d1}[E]_0}{(k_{d2} - k_{d1})} (e^{(-k_{d1}t)} - e^{(-k_{d2}t)}) \tag{1.29}$$

If the term fractional activity (A^f) is introduced where:

$$A^f = ([E] + A_1[E_1] + A_2[E_2])/[E]_0 \tag{1.30}$$

then, substituting for [E_2] from equation (1.24), gives:

$$A^f = \{[E] + A_1[E_1] + A_2([E]_0 - [E] - [E_1])\}/[E]_0 \tag{1.31}$$

therefore:

$$A^f = A_2 + \left\{1 + \frac{(A_1 k_{d1} - A_2 k_{d2})}{k_{d2} - k_{d1}}\right\} e^{(-k_{d1}t)} - \frac{(A_1 - A_2)k_{d1}}{k_{d2} - k_{d1}} e^{(-k_{d2}t)} \tag{1.32}$$

When both A_1 and A_2 are zero, the simple first-order deactivation rate expression results:

$$A^f = e^{(-k_{d1}t)} \tag{1.33}$$

The *half-life* $(t_{1/2})$ of an enzyme is the time it takes for the activity to reduce to a half of the original activity (i.e. $A^f = 0.5$). If the enzyme inactivation obeys equation (1.33), the half-life may be simply derived:

$$\ln(0.5) = - k_{d1}t_{1/2} \tag{1.34}$$

therefore:

$$t_{1/2} = 0.693/k_{d1} \tag{1.35}$$

In this simple case, the half-life of the enzyme is shown to be inversely proportional to the rate of denaturation.

Many enzyme preparations, both free and immobilised, appear to follow this series-type deactivation scheme. However, because reliable and reproducible data are difficult to obtain, inactivation data should, in general, be assumed to be rather error prone. It is not surprising that such data can be made to fit a model involving four determined parameters (A_1, A_2, k_{d1} and k_{d2}). Despite this possible reservation, equations (1.32) and (1.33) remain quite useful and the theory possesses the definite advantage of simplicity. In some cases the series-type deactivation may be due to structural microheterogeneity, where the enzyme preparation consists of a mixture of a large number of closely related structural forms. These may have been formed during the past history of the enzyme during preparation and storage, due to a number of minor reactions such as deamidation of one or two asparagine or glutamine residues, limited proteolysis or disulphide interchange. Alternatively it may be due to quaternary structure equilibria or the presence of distinct genetic variants. In any case, the larger the variability the more apparent will be the series-type inactivation kinetics. The practical effect of this is that usually k_{d1} is apparently much larger than k_{d2} and A_1 is less than unity.

In order to minimise loss of activity on storage, even moderate temperatures should be avoided. Most enzymes are stable for months if refrigerated (0–4 °C). Cooling below 0 °C, in the presence of additives (e.g. glycerol) which prevent freezing, can generally increase this storage stability even further. Freezing enzyme solutions is best avoided, as it often causes denaturation due to stress and pH variation caused by ice-crystal formation. The first-order deactivation constants are often significantly lower in the case of enzyme–substrate, enzyme–inhibitor and enzyme–product complexes, which helps to explain the substantial stabilising effects of suitable ligands, especially at concentrations where little free enzyme exists (e.g. $[S] \gg K_m$). Other factors, such as the presence of thiol anti-oxidants, may improve the thermal stability in particular cases.

It has been found that heat denaturation of enzymes is primarily due to the interactions of proteins with the aqueous environment. Proteins are generally more stable in concentrated, rather than dilute, solutions. In a dry or predominantly dehydrated state they remain active for considerable periods, even at temperatures above 100 °C. This property has great technological significance and is currently being exploited by the use of organic solvents (see Chapter 7).

Pressure changes will also affect enzyme-catalysed reactions. Clearly any reaction involving dissolved gases (e.g. oxygenases and decarboxylases) will be particularly affected by increased gas solubility at high pressures. The equilibrium position of the reaction will also be shifted due to any difference in molar volumes between the reactants and products. However an additional, if rather small, influence is due to the volume changes which occur during enzymic binding and catalysis. Some enzyme–reactant mixtures may undergo reductions in volume amounting to up to 50 ml mol^{-1} during reaction, due to conformational restrictions and changes in their hydration. This in turn, may lead to a doubling of the k_{cat}, and/or a halving in the K_m for a 1000-fold increase in pressure. The relative effects on k_{cat} and K_m depend upon the relative volume changes during binding and the formation of the reaction transition states.

Reversible reactions

A reversible enzymic reaction (e.g. the conversion of D-glucose to D-fructose, catalysed by glucose isomerase) may be represented by the following scheme where the reaction goes through the reversible stages of enzyme–substrate complex formation, conversion to enzyme–product complex and finally desorption of the product. No step is completely rate controlling:

$$E + S \underset{k_{-1}}{\overset{k_{+1}}{\rightleftharpoons}} ES \underset{k_{-2}}{\overset{k_{+2}}{\rightleftharpoons}} EP \underset{k_{-3}}{\overset{k_{+3}}{\rightleftharpoons}} E + P \qquad [1.12]$$

Pairs of symmetrical equations may be obtained for the change in the concentration of the intermediates with time:

$$d[ES]/dt = k_{+1}[E][S] + k_{-2}[EP] - (k_{-1} + k_{+2})[ES] \qquad (1.36)$$

$$d[EP]/dt = k_{-3}[E][P] + k_{+2}[ES] - (k_{+3} + k_{-2})[EP] \qquad (1.37)$$

Assuming that there is no denaturation, the total enzyme concentration must remain constant and:

$$[E] + [ES] + [EP] = [E]_0 \qquad (1.38)$$

therefore:

$$d[ES]/dt = k_{+1}([E]_0 - [EP])[S] + k_{-2}[EP]$$
$$- (k_{+1}[S] + k_{-1} + k_{+2})[ES] \qquad (1.39)$$

and:

$$d[EP]/dt = k_{-3}([E]_0 - [ES])[P] + k_{+2}[ES]$$
$$- (k_{-3}[P] + k_{+3} + k_{-2})[EP] \qquad (1.40)$$

Under conditions similar to those discussed earlier for the Michaelis–Menten mechanism (e.g. under the steady-state assumptions when both $d[ES]/dt$ and $d[EP]/dt$ are zero):

$$\frac{d[ES]/dt}{k_{+1}[S] + k_{-1} + k_{+2}} \ll [ES] \qquad (1.41)$$

and:

$$\frac{d[EP]/dt}{k_{-3}[P] + k_{+3} + k_{-2}} \ll [EP] \qquad (1.42)$$

are both true. The following symmetrical equations may be derived from equations (1.39) and (1.40) by using the approximations given by equations (1.41) and (1.42), and collecting terms:

$$[ES] = \frac{k_{-3}k_{-2}[E]_0[P] + (k_{+3} + k_{-2})k_{+1}[E]_0[S]}{(k_{+3}k_{+2} + k_{+3}k_{-1} + k_{-2}k_{-1}) + (k_{-2} + k_{+2} + k_{-1})k_{-3}[P] + (k_{+3} + k_{-2} + k_{+2})k_{+1}[S]} \qquad (1.43)$$

$$[EP] = \frac{k_{+1}k_{+2}[E]_0[S] + (k_{-1} + k_{+2})k_{-3}[E]_0[P]}{(k_{-1}k_{-2} + k_{-1}k_{+3} + k_{+2}k_{+3}) + (k_{+2} + k_{-2} + k_{+3})k_{+1}[S] + (k_{-1} + k_{+2} + k_{-2})k_{-3}[P]} \qquad (1.44)$$

The net rate of reaction (i.e. rate at which substrate is converted to product less the rate at which product is converted to substrate) may be denoted by v where:

$$v = k_{+2}[ES] - k_{-2}[EP] \qquad (1.45)$$

Substituting from equations (1.43) and (1.44):

$$v = \frac{k_{+1}k_{+2}k_{+3}[E]_0[S] - k_{-1}k_{-2}k_{-3}[E]_0[P]}{(k_{-1}k_{-2} + k_{-1}k_{+3} + k_{+2}k_{+3}) + (k_{+2} + k_{-2} + k_{+3})k_{+1}[S] + (k_{-2} + k_{+2} + k_{-1})k_{-3}[P]} \qquad (1.46)$$

therefore:

$$v = \frac{V^f[S]/K_m^S - V^r[P]/K_m^P}{1 + [S]/K_m^S + [P]/K_m^P} \qquad (1.47)$$

where:

$$V^f = \frac{k_{+2}k_{+3}[E]_0}{k_{+2} + k_{-2} + k_{+3}} \qquad (1.48)$$

$$V^r = \frac{k_{-2}k_{-1}[E]_0}{k_{-2} + k_{+2} + k_{-1}} \tag{1.49}$$

$$K_m^S = \frac{k_{-1}k_{-2} + k_{-1}k_{+3} + k_{+2}k_{+3}}{k_{+1}(k_{+2} + k_{-2} + k_{+3})} \tag{1.50}$$

$$K_m^P = \frac{k_{+3}k_{+2} + k_{+3}k_{-1} + k_{-2}k_{-1}}{k_{-3}(k_{-2} + k_{+2} + k_{-1})} \tag{1.51}$$

At equilibirum:

$$v = 0 \tag{1.52}$$

and, because the numerator of equation (1.47) must equal zero:

$$V^f[S]_\infty/K_m^S = V^r[P]_\infty/K_m^P \tag{1.53}$$

where $[S]_\infty$ and $[P]_\infty$ are the equilibrium concentrations of substrate and product (at infinite time). But by definition:

$$K_{eq} = [P]_\infty/[S]_\infty \tag{1.54}$$

Substituting from equation (1.53),

$$K_{eq} = V^f K_m^P / V^r K_m^S \tag{1.55}$$

This is the *Haldane* relationship.

Therefore:

$$v = \frac{(V^f/K_m^S)([S] - [P]/K_{eq})}{1 + [S]/K_m^S + [P]/K_m^P} \tag{1.56}$$

If K_m^S and K_m^P are approximately equal (e.g. the commercial immobilised glucose isomerase, Sweetase, has K_m (glucose) of 840 mM and K_m (fructose) of 830 mM at 70 °C, and noting that the total amount of substrate and product at any time must equal the sum of the substrate and product at the start of the reaction:

$$[S] + [P] = [S]_0 + [P]_0 \tag{1.57}$$

$$v = \frac{V^f([S] - [P]/K_{eq})}{K_m^S + [S]_0 + [P]_0} \tag{1.58}$$

therefore:

$$v = K'([S] - [P]/K_{eq}) \tag{1.59}$$

where:

$$K' = \frac{V^f}{K_m^S + [S]_0 + [P]_0} \tag{1.60}$$

K' is not a true kinetic constant as it is only constant if the initial substrate plus product concentration is kept constant.

Also:

$$[S] + [P] = [S]_\infty + [P]_\infty \tag{1.61}$$

Substituting from equation (1.54):

$$[P] = K_{eq}[S]_\infty + [S]_\infty - [S] \tag{1.62}$$

Let $[S^*]$ equal the concentration difference between the actual concentration of substrate and the equilibrium concentration:

$$[S^*] = [S] - [S]_\infty \tag{1.63}$$

therefore:

$$[P] = K_{eq}[S]_\infty - [S^*] \tag{1.64}$$

Substituting in equation (1.47):

$$v = \frac{V^f[S^*]/K_m^S + V^f[S]_\infty/K_m^S - V^r K_{eq}[S]_\infty/K_m^P + V^f[S^*]/K_m^P}{1 + [S^*]/K_m^S + [S]_\infty/K_m^S + K_{eq}[S]_\infty/K_m^P - [S^*]/K_m^P} \tag{1.65}$$

Rearranging equation (1.55):

$$V^f/K_m^S = V^r K_{eq}/K_m^P \tag{1.66}$$

therefore:

$$v = \frac{(V^f/K_m^S + V^r/K_m^P)[S^*]}{1 + [S]_\infty/K_m^S + K_{eq}[S]_\infty/K_m^P + \dfrac{(K_m^P - K_m^S)}{K_m^S K_m^P}[S^*]} \tag{1.67}$$

therefore:

$$v = \frac{V[S^*]}{K + [S^*]} \tag{1.68}$$

where:

$$V = \frac{K_m^P V^f + K_m^S V^r}{K_m^P - K_m^S} \tag{1.69}$$

therefore:

$$V = \frac{K_m^P V^f + K_m^P V^f/K_{eq}}{K_m^P - K_m^S} \tag{1.70}$$

therefore:

$$V = \frac{(K_{eq} + 1)}{K_{eq}} \times \frac{K_m^P}{(K_m^P - K_m^S)} \times V^f \tag{1.71}$$

and:

$$K = \frac{K_m^S K_m^P + (K_m^P + K_m^S K_{eq})[S]_\infty}{K_m^P - K_m^S} \tag{1.72}$$

As in the case of K' in equation (1.59), K is not a true kinetic constant as it varies with $[S]_\infty$ and hence the sum of $[S]_0$ and $[P]_0$. It is only constant if the initial substrate plus product concentration is kept constant. By a similar but symmetrical argument, the net reverse rate of reaction:

$$v_{rev} = \frac{V_{rev}[P^*]}{K_{rev} + [P^*]}$$ (1.73)

with constants defined as above but by symmetrically exchanging K_m^P with K_m^S, and V^r with V^f.

Both equations (1.59) and (1.68) are useful when modelling reversible reactions, particularly the technologically important reaction catalysed by glucose isomerase. They may be developed further to give productivity–time estimates and for use in the comparison of different reactor configurations (see Chapters 3 and 5).

Although an enzyme can never change the equilibrium position of a catalysed reaction, as it has no effect on the standard free energy change involved, it can favour reaction in one direction rather than its reverse. It achieves this by binding the reactants strongly, as enzyme–reactant complexes, in this preferred direction but binding the product(s) only weakly. The enzyme is bound up with the reactant(s), encouraging their reaction, leaving little enzyme free to catalyse the reaction in the reverse direction. It is unlikely, therefore, that the same enzyme preparation would be optimum for catalysing a reversible reaction in both directions.

Enzyme inhibition

A number of substances may cause a reduction in the rate of an enzyme-catalysed reaction. Some of these (e.g. urea) are non-specific protein denaturants. Others, which act generally in a fairly specific manner, are known as inhibitors. Loss of activity may be either reversible, where activity may be restored by the removal of the inhibitor, or irreversible, where the loss of activity is time dependent and cannot be recovered during the time-scale of interest. If the inhibited enzyme is totally inactive, irreversible inhibition behaves as a time-dependent loss of enzyme concentration (i.e. lower V_{max}), in other cases, involving incomplete inactivation, there may be time-dependent changes in both K_m and V_{max}. Heavy-metal ions (e.g. mercury and lead) should generally be prevented from coming into contact with enzymes as they usually cause irreversible inhibition by binding strongly to the amino acid backbone.

More important for most enzyme-catalysed processes is the effect of reversible inhibitors. These are generally discussed in terms of a simple extension to the Michaelis–Menten reaction scheme.

[1.13]

where I represents the reversible inhibitor and the inhibitory (dissociation) constants K_i and K_i' are given by:

$$K_i = \frac{[E][I]}{[EI]} \tag{1.74}$$

and:

$$K_i' = \frac{[ES][I]}{[ESI]} \tag{1.75}$$

For the present purposes, it is assumed that neither EI nor ESI may react to form product. Equilibrium between EI and ESI is allowed, but makes no net contribution to the rate equation as it must be equivalent to the equilibrium established through:

$$EI + S \rightleftharpoons E + S + I \rightleftharpoons ES + I \rightleftharpoons ESI \tag{1.14}$$

Binding of inhibitors may change with the pH of the solution, as discussed earlier for substrate binding, and result in the independent variation of both K_i and K_i' with pH.

In order to simplify the analysis substantially, it is necessary that the rate of product formation (k_{+2}) is slow relative to the establishment of the equilibria between the species.
Therefore:

$$K_m = k_{-1}/k_{+1} = [E][S]/[ES] \tag{1.76}$$

also:

$$\frac{v}{V_{max}} = \frac{[ES]}{[E]_0} \tag{1.77}$$

where:

$$[E]_0 = [E] + [EI] + [ES] + [ESI] \tag{1.78}$$

therefore:

$$\frac{v}{V_{max}} = \frac{[ES]}{[E] + [EI] + [ES] + [ESI]} \tag{1.79}$$

Substituting from equations (1.74), (1.75) and (1.76), followed by simplification, gives:

$$\frac{v}{V_{max}} = \frac{1}{\dfrac{K_m}{[S]} + \dfrac{K_m[I]}{[S]K_i} + 1 + \dfrac{[I]}{K_i'}} \tag{1.80}$$

therefore:

$$v = \frac{V_{max}[S]}{K_m(1 + [I]/K_i) + [S](1 + [I]/K_i')} \tag{1.81}$$

If the total enzyme concentration is much less than the total inhibitor concentration (i.e. $[E]_0 \ll [I]_0$), then:

$$v = \frac{V_{max}[S]}{K_m(1 + [I]_0/K_i) + [S](1 + [I]_0/K_i')} \tag{1.82}$$

This is the equation used generally for *mixed inhibition* involving both EI and ESI complexes (Figure 1.8(*a*)). A number of simplified cases exist.

Competitive inhibition

K_i' is much greater than the total inhibitor concentration and the ESI complex is not formed. This occurs when both the substrate and inhibitor compete for binding to the active site of the enzyme. The inhibition is most noticeable at low substrate concentrations but can be overcome at sufficiently high substrate concentrations, as the V_{max} remains unaffected (Figure 1.8(*b*)). The rate equation is given by:

$$v = V_{max}[S]/(K_m^{app} + [S]) \tag{1.83}$$

where K_m^{app} is the apparent K_m for the reaction, and is given by:

$$K_m^{app} = K_m(1 + [I]/K_i) \tag{1.84}$$

Normally the competitive inhibitor bears some structural similarity to the substrate, and often is a reaction product (*product inhibition*, e.g. inhibition of lactase by galactose), which may cause a substantial loss of productivity when high degrees of conversion are required. The rate equation for product inhibition is derived from equations (1.83) and (1.84):

$$v = \frac{V_{max}[S]}{K_m(1 + [P]/K_P) + [S]} \tag{1.85}$$

A similar effect is observed with competing substrates, quite a common state of affairs in industrial conversions, and especially relevant to macromolecular hydrolyses where a number of different substrates may coexist, all with different kinetic parameters. The reaction involving two co-substrates may be modelled by the scheme:

$$E + S_1 \underset{k_{-1}}{\overset{k_{+1}}{\rightleftharpoons}} ES_1 \xrightarrow{k_{+2}} P_1$$

$$k_{-3} \Big\Updownarrow k_{+3} \quad +S_2$$

$$ES_2 \xrightarrow{k_{+4}} P_2$$

[1.15]

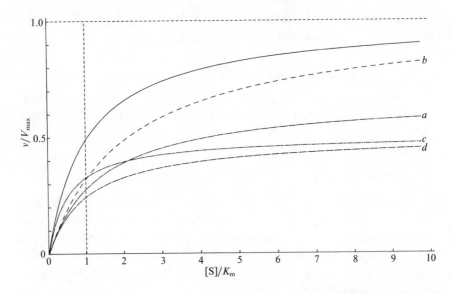

Figure 1.8. Diagram showing the effect of reversible inhibitors on the rate of enzyme-catalysed reactions. —— no inhibition, (*a*) −−−−− mixed inhibition ($[I] = K_i = 0.5\ K_i'$); lower V_{max}^{app} ($= 0.67\ V_{max}$), higher K_m^{app} ($= 2\ K_m$). (*b*) −−−−− competitive inhibition ($[I] = K_i$); V_{max}^{app} unchanged ($= V_{max}$), higher K_m^{app} ($= 2\ K_m$). (*c*) −··−··−·· uncompetitive inhibition ($[I] = K_i'$); lower V_{max}^{app} ($= 0.5\ V_{max}$) and K_m^{app} ($= 0.5\ K_m$). (*d*) −·−·−· non-competitive inhibition ($[I] = K_i = K_i'$); lower V_{max}^{app} ($= 0.5\ V_{max}$), unchanged K_m^{app} ($= K_m$).

Both substrates compete for the same catalytic site and, therefore, their binding is mutually exclusive and they behave as competitive inhibitors of each others' reactions. If the rates of product formation are much slower than attainment of the equilibria (i.e. k_{+2} and k_{+4} are very much less than k_{-1} and k_{-3}, respectively), the rate of formation of P_1 is given by:

$$v_1 = \frac{V_{max}^1[S_1]}{K_m^1(1 + [S_2]/K_m^2) + [S_1]} \tag{1.86}$$

and the rate of formation of P_2 is given by:

$$v_2 = \frac{V_{max}^2[S_2]}{K_m^2(1 + [S_1]/K_m^1) + [S_2]} \tag{1.87}$$

If the substrate concentrations are both small relative to their K_m values:

$$\frac{v_1}{v_2} = \frac{V_{max}^1/K_m^1}{V_{max}^2/K_m^2} \times \frac{[S_1]}{[S_2]} \tag{1.88}$$

Therefore, in a competitive situation using the same enzyme and with both substrates at the same concentration:

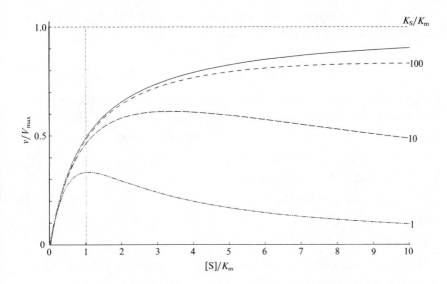

Figure 1.9. The effect of substrate inhibition on the rate of an enzyme-catalysed reaction. A comparison is made between the inhibition caused by increasing K_S relative to K_m: —— no inhibition, $K_s/K_m \gg 100$; ------ $K_s/K_m = 100$; —·—·—·. $K_s/K_m = 10$; —···—···—·· $K_s/K_m = 1$. By the nature of the binding causing this inhibition, it is unlikely that $K_s/K_m < 1$.

$$\frac{v_1}{v_2} = \frac{k_{cat}^1/K_m^1}{k_{cat}^2/K_m^2} \qquad (1.89)$$

where $k_{cat}^1 = k_{+2}$ and $k_{cat}^2 = k_{+4}$ in this simplified case. The relative rates of reaction are in the ratio of their specificity constants. If both reactions produce the same product (e.g. some hydrolyses):

$$v = \frac{V_{max}^1[S_1]}{K_m^1(1 + [S_2]/K_m^2 + [S_1]/K_m^1)} + \frac{V_{max}^2[S_2]}{K_m^2(1 + [S_1]/K_m^1 + [S_2]/K_m^2)} \qquad (1.90)$$

therefore:

$$v = \frac{V_{max}^1[S_1]/K_m^1 + V_{max}^2[S_2]/K_m^2}{1 + [S_1]/K_m^1 + [S_2]/K_m^2} \qquad (1.91)$$

Uncompetitive inhibition

K_i is much greater than the total inhibitor concentration and the EI complex is not formed. This occurs when the inhibitor binds to a site which only becomes available after the substrate (S_1) has bound to the active site of the enzyme. This inhibition is most commonly encountered in multi substrate reactions where the inhibitor is competitive with respect to one substrate (e.g. S_2) but uncompetitive wiith respect to another (e.g. S_1), where the reaction scheme may be represented by:

$$E + S_1 \rightleftharpoons ES_1 + S_2 \rightleftharpoons ES_1S_2 \longrightarrow$$
$$E + S_1 \rightleftharpoons ES_1 + I \rightleftharpoons ES_1I \;—X— \qquad \text{products} \qquad [1.16]$$

The inhibition is most noticeable at high concentrations of substrate (i.e. S_1 in the scheme above) and cannot be overcome as both the V_{max} and K_m are equally reduced (Figure 1.8, curve c). The rate equation is:

$$v = V_{max}^{app}[S]/(K_m^{app} + [S]) \qquad (1.92)$$

where V_{max}^{app} and K_m^{app} are the apparent V_{max} and K_m given by:

$$V_{max}^{app} = V_{max}/(1 + [I]/K_i') \qquad (1.93)$$

and:

$$K_m^{app} = K_m/(1 + [I]/K_i') \qquad (1.94)$$

In this case the specificity constant remains unaffected by the inhibition. Normally the uncompetitive inhibitor also bears some structural similarity to one of the substrates and, again, is often a reaction product.

A special case of uncompetitive inhibition is *substrate inhibition*, which occurs at high concentrations of substrate in about 20% of all known enzymes (e.g. invertase is inhibited by sucrose). It is caused primarily by more than one substrate molecule binding to an active site meant for just one, often different parts of the substrate molecules binding to different subsites within the substrate-binding site. If the resultant complex is inactive this type of inhibition causes a reduction in the rate of reaction, at high concentrations of substrate. It may be modelled by the following scheme:

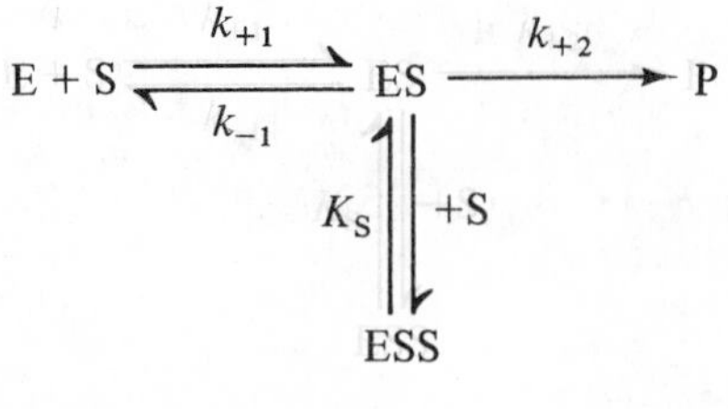

$$E + S \underset{k_{-1}}{\overset{k_{+1}}{\rightleftharpoons}} ES \xrightarrow{k_{+2}} P$$
$$K_S \Big\updownarrow +S \qquad\qquad [1.17]$$
$$ESS$$

where:

$$K_s = \frac{[ES][S]}{[ESS]} \qquad (1.95)$$

The assumption is made that ESS may not react to form product. It follows from equation (1.82) that:

$$v = \frac{V_{max}[S]}{K_m + [S](1 + [S]/K_S)} \qquad (1.96)$$

Even quite high values for K_s lead to a levelling off of the rate of reaction at

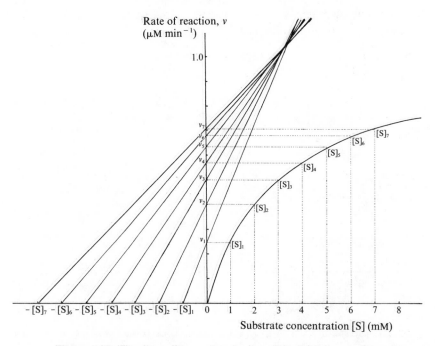

Figure 1.10. The direct linear plot. A plot of the initial rate of reaction against the initial substrate concentration also showing the way estimates can be made directly of the K_m and V_{max}. Every pair of data points may be utilised to give a separate estimate of these parameters (i.e. $n(n-1)/2$ estimates from n data points with differing $[S]_0$). These estimates are determined from the intersections of lines passing through the (x, y) points $(-[S]_0, 0)$ and $(0,v)$; each intersection forming a separate estimate of K_m and V_{max}. The intersections are separately ranked in order of increasing value of both K_m and V_{max} and the median values taken as the best estimates for these parameters. The error in these estimates can be simply determined from subranges of these estimates, the width of the subrange being dependent on the accuracy required for the error and the number of data points in the analysis. In this example there are seven data points and, therefore, 21 estimates for both K_m and V_{max}. The ranked list of the estimates for K_m (mM) is 1.9, 2.6, 2.7, 2.8, 2.8, 2.8, 3.0, 3.0, 3.0, 3.0, 3.0, 3.0, 3.1, 3.1, 3.1, 3.2, 3.2, 3.3, 3.3, 3.3, 3.7, with a median value of 3.0 mM. The K_m must lie between the fourth (2.8mM) and eighteenth (3.3 mM) estimates at a confidence level of 97% (Cornish-Bowden *et al.*, 1978). It can be seen that outlying estimates have little or no influence on the results. This is a major advantage over the least-squared statistical procedures where rogue data points cause heavily biased effects.

(a)

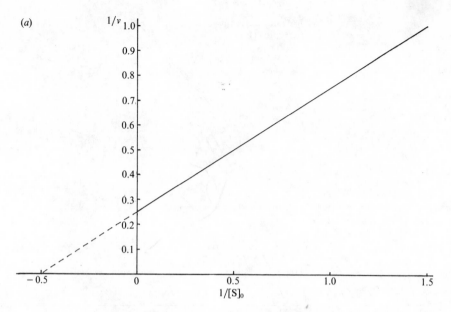

(b)

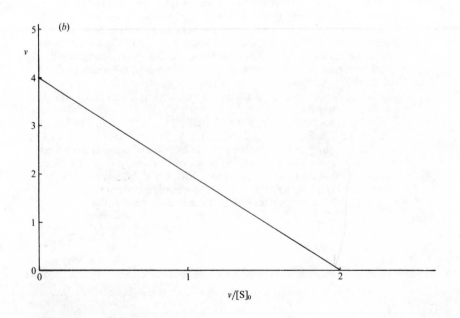

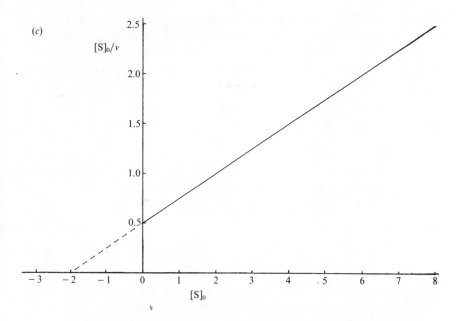

Figure 1.11. Three ways in which the hyperbolic relationship between the initial rate of reaction and the initial substrate concentration ($v = V_{max}[S]_0/(K_m + [S]_0)$) can be rearranged to give linear plots. The examples are drawn using $K_m = 2$ mM and $V_{max} = 4$ μM min^{-1}.
(*a*) Lineweaver–Burk (double-reciprocal) plot of $1/v$ against $1/[S]_0$ giving intercepts at $1/V_{max}$ and $-1/K_m$:

$$\frac{1}{v} = \frac{K_m}{V_{max}} \times \frac{1}{[S]_0} + \frac{1}{V_{max}} \qquad (1.103)$$

(*b*) Eadie–Hofstee plot of v against $v/[S]_0$ giving intercepts at V_{max} and V_{max}/K_m:

$$v = -K_m \times \frac{v}{[S]_0} + V_{max} \qquad (1.104)$$

(*c*) Hanes–Woolf (half-reciprocal) plot of $[S]_0/v$ against $[S]_0$ giving intercepts at K_m/V_{max} and K_m:

$$\frac{[S]_0}{v} = \frac{1}{V_{max}} \times [S]_0 + \frac{K_m}{V_{max}} \qquad (1.105)$$

high substrate concentrations, and lower K_S values cause substantial inhibition (Figure 1.9).

Non-competitive inhibition

Both the EI and ESI complexes are formed equally well (i.e. $K_i = K_i'$). This occurs when the inhibitor binds at a site away from the substrate-binding site, causing a reduction in the catalytic rate. It is quite rarely found as a special case of mixed inhibition. The fractional inhibition is identical at all substrate concentrations and cannot be overcome by increasing substrate concentration due to the reduction in V_{max} (Figure 1.8, curve d). The rate equation is given by:

$$v = V_{max}^{app}[S]/(K_m + [S])$$ (1.97)

where V_{max}^{app} is given by:

$$V_{max}^{app} = V_{max}/(1 + [I]/K_i]$$ (1.98)

The diminution in the rate of reaction with pH, described earlier, may be considered as a special case of non-competitive inhibition, the inhibitor being the hydrogen ion on the acid side of the optimum or the hydroxide ion on the alkaline side.

Determination of V_{max} and K_m

It is important to have as thorough a knowledge as possible of the performance characteristics of enzymes, if they are to be used most efficiently. The kinetic parameters V_{max}, K_m and k_{cat}/K_m should, therefore, be determined. There are two approaches to this problem, using either the reaction progress curve (integral method) or the initial rates of reaction (differential method). Use of either method depends on prior knowledge of the mechanism for the reaction and, at least approximately, the optimum conditions for the reaction. If the mechanism is known and complex then the data must be reconciled to the appropriate model (hypothesis), usually by use of a computer-aided analysis involving a weighted least-squares fit. Many such computer programs are currently available and, if not, the programming skill involved is usually fairly low. If the mechanism is not known, initial attempts are usually made to fit the data to the Michaelis–Menten kinetic model. Combining equations (1.1) and (1.8):

$$\frac{d[P]}{dt} = v = \frac{V_{max}[S]}{K_m + [S]}$$ (1.99)

which, on integration, using the boundary condition that the product is absent at time zero and by substituting [S] by ([S]$_0$ – [P]), becomes:

$$t = \frac{[P]}{V_{max}} - \frac{K_m}{V_{max}}\ln\left(\frac{[S]_0 - [P]}{[S]_0}\right)$$ (1.100)

If the fractional conversion (X) is introduced, where

$$X = ([S]_0 - [S])/[S]_0$$ (1.101)

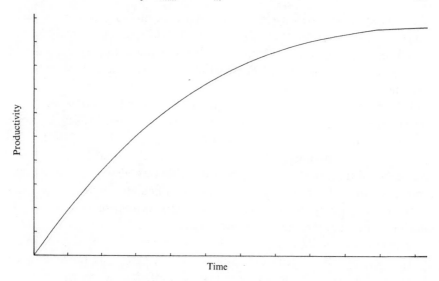

Figure 1.12. A schematic plot showing the amount of product formed (productivity) against the time of reaction, in a closed system. The specificity constant may be determined by a weighted least-squares fit of the data to the relationship given by equation (1.102).

then equation (1.100) may be simplified to give:

$$t = \frac{X[S]_0}{V_{max}} - \frac{K_m \ln(1 - X)}{V_{max}} \qquad (1.102)$$

Use of equation (1.99) involves the determination of the initial rate of reaction over a wide range of substrate concentrations. The initial rates are used so that $[S] = [S]_0$, the predetermined and accurately known substrate concentration at the start of the reaction. Its use also ensures that there is no effect of reaction reversibility or product inhibition which may affect the integral method based on equation (1.102). Equation (1.99) can be utilised directly using a computer program, involving a weighted least-squares fit, where the parameters for determining the hyperbolic relationship between the initial rate of reaction and initial substrate concentration (i.e. K_m and V_{max}) are chosen in order to minimise the errors between the data and the model, and the assumption is made that the errors inherent in the practically determined data are normally distributed about their mean (error-free) value.

Alternatively, the direct linear plot may be used (Figure 1.10). This is a powerful non-parametric statistical method which depends upon the assumption that any errors in the experimentally derived data are as likely to be positive (i.e. too high) as negative (i.e. too low). It is common practice to show the data obtained by the above statistical methods on one of three linearised plots, derived from equation (1.99) (Figure 1.11). Of these, the double reciprocal plot is preferred to test for the qualitative correctness of a proposed mechanism, and the Eadie–Hofstee plot is preferred for discovering deviations from linearity.

The progress curve of the reaction (Figure 1.12) can be used to determine the specificity constant (k_{cat}/K_m) by making use of the relationship between time of reaction and fractional conversion (see equation (1.102). This has the advantage over the use of the initial rates (above) in that fewer determinations need to be made, possibly only one progress curve is necessary, and sometimes the initial rate of reaction is rather difficult to determine due to its rapid decline. If only the early part of the progress curve, or its derivative, is utilised in the analysis, this procedure may even be used in cases where there is competitive inhibition by the product, or where the reaction is reversible.

The type of inhibition and the inhibition constants may be determined from the effect of differing concentrations of inhibitor on the apparent K_m, V_{max} and k_{cat}/K_m, although some more specialised plots do exist (see e.g. Cornish-Bowden, 1974).

Summary

(*a*) Enzymes are specific catalysts of vast range and utility.

(*b*) Their activity is governed by their structure and physical environment.

(*c*) Care should be taken over the interpretation of reported units of enzymic activity and the conditions necessary for maximum productivity.

(*d*) Enzymes may lose their catalytic activity reversibly or irreversibly due to denaturation or inhibition, depending upon the conditions.

(*e*) The values of the K_m, V_{max}, specificity constants, $pH_{optimum}$ and rate of thermal denaturation are all of relevance and utility to enzyme technologists.

Bibliography

Bender, M. L., Kezdy, J. F. & Gunter, C. R. (1964). The anatomy of an enzymatic catalysis: α-chymotrypsin. *Journal of the American Chemical Society*, **86**, 3714–21.

Chaplin, M. F. (1986). *Protein structure and enzyme activity*. Oxford: IRL Press Ltd. [This is a CAL/simulation software package suitable for IBM or BBC microcomputers.]

Cornish-Bowden, A. (1974). A simple graphical method for determining the inhibition constants of mixed, uncompetitive and non-competitive inhibitors. *Biochemical Journal*, **137**, 143–4.

Cornish-Bowden, A. (1976). *Principles of enzyme kinetics*. London: Butterworth.

Cornish-Bowden, A. & Endrenyi, L. (1986). Robust regression of enzyme kinetic data. *Biochemical Journal*, **234**, 21–9.

Cornish-Bowden, A., Porter, W. R. & Trager, W. F. (1978). Evaluation of distribution-free confidence limits for enzyme kinetic parameters. *Journal of Theoretical Biology*, **74**, 163–75.

Crompton, I. E. & Waley, S. G. (1986). The determination of specificity constants in enzyme-catalysed reactions. *Biochemical Journal*, **239**, 221–4.

Eisenthal, R. & Cornish-Bowden, A. (1974). The direct linear plot. *Biochemical Journal*, **139**, 715–20.

Fersht, A. (1985). *Enzyme structure and mechanism*, 2nd edn. New York: W. H. Freeman & Co.

Henderson, P. J. F. (1978). Statistical analysis of enzyme kinetic data. *Techniques in the life sciences: biochemistry*, vol. B1/11 *Techniques in protein and enzyme biochemistry*, part 2, pp. B113/1–41. Amsterdam: Elsevier/North-Holland Biomedical Press.

Henley, J. P. & Sadana, A. (1985). Categorization of enzyme deactivations using a series-type mechanism. *Enzyme and Microbial Technology*, **7**, 50–60.

Hill, C. M., Waight, R. D. & Bardsley, W. G. (1977). Does any enzyme follow the Michaelis–Menten equation? *Molecular and Cellular Biochemistry*, **15**, 173–8.

Koshland, D. E., Jr (1962). The comparison of non-enzymic and enzymic reaction velocities. *Journal of Theoretical Biology*, **2**, 75–86.

Michaelis, L. & Menten, M. L. (1913). The kinetics of invertin action. *Biochemische Zeitschrift*, **49**, 333–69.

Nomenclature Committee of the International Union of Biochemistry, (1984). *Enzyme Nomenclature: recommendations (1984) of the Nomenclature Committee of the International Union of Biochemistry*. Orlando, FL: Academic Press.

Wong, J. T. (1965). On the steady-state method of enzyme kinetics. *Journal of the American Chemical Society*, **87**, 1788–93.

2   Enzyme preparation and use

Sources of enzymes

Biologically active enzymes may be extracted from any living organism. A very wide range of sources is used for commercial enzyme production – from *Actinoplanes* to *Zymomonas*, from spinach to snake venom. Of the hundred or so enzymes being used industrially, over half are from fungi and yeast and over a third are from bacteria, with the remainder divided between animal (8%) and plant (4%) sources (Table 2.1). A very much larger number of enzymes find uses in chemical analysis and clinical diagnosis. Non-microbial sources comprise a larger proportion of these, at the present time. Microbes are preferred to plants and animals as sources of enzymes because: (1) they are generally cheaper to produce; (2) their enzyme contents are more predictable and controllable; (3) reliable supplies of raw material of constant composition are more easily arranged; and (4) plant and animal tissues contain more potentially harmful materials, including phenolic compounds (from plants), endogenous enzyme inhibitors and proteases. Attempts are being made to overcome some of these difficulties by the use of animal and plant cell cultures.

In practice, the great majority of microbial enzymes come from a very limited number of genera, of which *Aspergillus* species, *Bacillus* species and *Kluyveromyces* (also called *Saccharomyces*) species predominate. Most of the strains used have either been employed by the food industry for many years or been derived from such strains by mutation and selection. There are very few instances of the industrial use of enzymes which have been developed for one task. Shining examples of such developments are the production of high-fructose syrup using glucose isomerase and the use of pullulanase in starch hydrolysis.

Producers of industrial enzymes and their customers will share the common aims of economy, effectiveness and safety. They will wish to have high-yielding strains of microbes which make the enzyme constitutively and secrete it into their growth medium (extracellular enzymes). If the enzyme is not produced constitutively, induction must be rapid and inexpensive.

Producers will aim to use strains of microbe that are known to be generally safe. Users will pay little regard to the way in which the enzyme is produced but will insist on having preparations that have a known activity and keep that activity for extended periods, stored at room temperature or with routine refrigeration. They will pay little attention to the purity of the enzyme preparation provided that it does not contain materials (enzymes or not) that interfere with their process. Both producers and users will wish to have the enzymes in forms that present minimal hazard to those handling them or to those consuming the product.

The development of commercial enzymes is a specialised business which is usually undertaken by a handful of companies that have high skills in (1) screening for new and improved enzymes, (2) fermentation for enzyme production, (3) large-scale enzyme purifications, (4) formulation of enzymes for sale, (5) customer liaison, and (6) dealing with the regulatory authorities.

Screening for novel enzymes

If a reaction is thermodynamically possible, it is likely that an enzyme exists which is capable of catalysing it. One of the major skills of enzyme-producing companies and suitably funded academic laboratories is the rapid and cost-effective screening of microbial cultures for enzyme activities. Natural samples, usually soil or compost material found near high concentrations of likely substrates, are used as sources of cultures. It is not unusual at international congresses of enzyme technologists to see representatives of enzyme companies collecting samples of soil to be screened later when they return to their laboratories.

The first stage of the screening procedure for commercial enzymes is to screen ideas, i.e. to determine the potential commercial need for a new enzyme, to estimate the size of the market and to decide, approximately, how much potential users of the enzyme will be able to afford to pay for it. In some cases, the determination of the potential value of an enzyme is not easy, for instance when it might be used to produce an entirely novel substance. In others, for instance when the novel enzyme would be used to improve an existing process, its potential value can be costed very accurately. In either case, a cumulative cash flow must be estimated, balancing the initial screening and investment capital costs, including interest, tax liability and depreciation, against the expected long-term profits. Full account must be taken of inflation, projected variation in feedstock price and source, publicity and other costs. In addition, the probability of potential market competition and changes in political or legal factors must be considered. Usually the sensitivity of the project to changes in all of these factors must be estimated,

Table 2.1 *Some important industrial enzymes and their sources*

Enzyme[a]	EC number[b]	Source	Intra/extra-cellular[c]	Scale of production[d]	Industrial use
Animal enzymes					
Catalase	1.11.1.6	Liver	I	−	Food
Chymotrypsin	3.4.21.1	Pancreas	E	−	Leather
Lipase[e]	3.1.1.3	Pancreas	E	−	Food
Rennet[f]	3.4.23.4	Abomasum	E	+	Cheese
Trypsin	3.4.21.4	Pancreas	E	−	Leather
Plant enzymes					
Actinidin	3.4.22.14	Kiwi fruit	E	−	Food
α-Amylase	3.2.1.1	Malted barley	E	+ +	Brewing
β-Amylase	3.2.1.1.2	Malted barley	E	+ +	Brewing
Bromelain	3.4.22.4	Pineapple latex	E	−	Brewing
β-Glucanase[g]	3.2.1.6	Malted barley	E	+	Brewing
Ficin	3.4.22.3	Fig latex	E	−	Food
Lipoxygenase	1.13.11.12	Soybeans	I	−	Food
Papain	3.4.22.2	Pawpaw latex	E	+	Meat
Bacterial enzymes					
α-Amylase	3.2.1.1	*Bacillus*	E	+	Starch
β-Amylase	3.2.1.2	*Bacillus*	E	+	Starch
Asparaginase	3.5.1.1	*Escherichia coli*	I	−	Health
Glucose isomerase[h]	5.3.1.5	*Bacillus, Streptomyces*	I	+	Fructose syrup
Penicillin amidase	3.5.1.11	*Bacillus*	I	−	Pharmaceutical
Protease[i]	3.4.21.14	*Bacillus*	E	+ +	Detergent

Enzyme	EC number[b]	Organism	[c]	[d]	Application
Pullulanase[j]	3.2.1.41	Klebsiella,Bacillus	E	–	Starch
Fungal enzymes					
α-Amylase	3.2.1.1	Aspergillus	E	+ +	Baking
Aminoacylase	3.5.1.14	Aspergillus	I	–	Pharmaceutical
Glucoamylase[k]	3.2.1.3	Aspergillus,Rhizopus	E	+ + +	Starch
Catalase	1.11.1.6	Aspergillus	I	–	Food
Cellulase	3.2.1.4	Trichoderma	E	–	Waste
Dextranase	3.2.1.11	Pencillium	E	–	Food
Glucose oxidase	1.1.3.4	Aspergillus	I	–	Food
Lactase[l]	3.2.1.23	Aspergillus	E	–	Dairy
Lipase[e]	3.1.1.3	Rhizopus	E	–	Food
Rennet[m]	3.4.23.6	Mucor miehei	E	+ +	Cheese
Pectinase[n]	3.2.1.15	Aspergillus	E	+ +	Drinks
Pectin lyase	4.2.2.10	Aspergillus	E	–	Drinks
Protease[m]	3.4.23.6	Aspergillus	E	+	Baking
Raffinase[o]	3.2.1.22	Mortierella	I	–	Food
Yeast enzymes					
Invertase[p]	3.2.1.26	Saccharomyces	I/E	–	Confectionery
Lactase[l]	3.2.1.23	Kluyveromyces	I/E	–	Dairy
Lipase[e]	3.1.1.3	Candida	E	–	Food
Raffinase[o]	3.2.1.11	Saccharomyces	I	–	Food

[a] The names in common usage are given. As most industrial enzymes consist of mixtures of enzymes, these names may vary from the recommended names of their principal component. Where appropriate, the recommended name of this principal component is given below.

[b] The EC number of the principal component.

[c] I, intracellular enzyme; E, extracellular enzyme.

[d] + + + > 100 tonnes year^{-1}; + + > 10 tonnes year^{-1}; + > 1 tonne year^{-1}; – < 1 tonne year^{-1}.

[e] Triacylglycerol lipase; [f] chymosin; [g] endo-1,3(4)-β-glucanase; [h] xylose isomerase; [i] subtilisin; [j] α-dextrin endo-1,6-α-glucosidase; [k] glucan 1,4-α-glucosidase; [l] β-galactosidase; [m] microbial aspartic proteinase; [n] polygalacturonase; [o] α-galactosidase; [p] β-fructofuranosidase.

by informed guesswork, in order to assess the risk factor involved. Financial re-appraisal must be carried out frequently during the development process to check that it still constitutes an efficient use of resources.

If agreement is reached, probably after discussions with potential users, that experimental work would be commercially justifiable, the next stage involves the location of a source of the required enzyme. Laboratory work is expensive in manpower so clearly it is worth while using all available databases to search for mention of the enzyme in the academic and patents literature. Cultures may then be sought from any sources so revealed. Some preparations of commercial enzymes are quite rich sources of enzymes other than the enzyme which is being offered for sale; thus such preparations may be worth investigating as potential inexpensive sources.

If these first searches are successful, it is probably necessary to screen for new microbial strains capable of performing the transformation required. This should not be a 'blind' screen: there will usually be some source of microbes that could have been exposed for countless generations to the conditions that the new enzyme should withstand or to chemicals which it is required to modify. Hence, thermophiles are sought in hot springs, osmophiles in sugar factories, organisms capable of metabolising wood preservatives in timber yards and so on. A classic example of the detection of an enzyme by intelligent screening was the discovery of a commercially useful cyanide-degrading enzyme in the microbial pathogens of plants that contain cyanogenic glycosides.

The identification of a microbial source of an enzyme is by no means the end of the story. The properties of the enzyme must be determined: temperature for optimum productivity, temperature stability profile, pH optimum and stability, kinetic constants (K_m, V_{max}), whether there is substrate or product inhibition, and the ability to withstand components of the expected feedstock other than substrate. A team of scientists, engineers and accountants must then consider the next steps. If any of these parameters is unsatisfactory, the screen must continue until improved enzymes are located. Now that protein engineering (see Chapter 8) can be contemplated seriously, an enzyme with sufficient potential value could be improved 'by design' to overcome one or two shortcomings. However, this would take a long time, at the present level of knowledge and skill, so further screening of microbes from selected sources should probably be considered more worth while.

Once an enzyme with suitable properties has been located, various decisions must be made concerning the acceptability of the organism to the regulatory authorities, the productivity of the organism, and the way in which the enzyme is to be isolated, utilised (free or immobilised) and, if necessary, purified. If the organism is unacceptable from a regulatory viewpoint, two options exist: to eliminate that organism altogether and continue the screening operation, or to clone the enzyme into an acceptable organism. The latter approach is becoming increasingly attractive, especially, as cloning could also be used to increase the productivity of the fermentation process. Cloning may also be attractive when the organism originally producing the enzyme is acceptable from the health and safety point of view but whose productivity is unacceptable (see Chapter 8). However, cloning is not yet routine and invariably successful so there is still an excellent case to be made for applying conventional mutation and isolation techniques for the selection of improved strains. It should be noted that, although the technology for cloning glucose isomerase into 'routine' organisms is known, it has not yet been applied. Several of the glucose isomerase preparations used commercially consist of whole cells, or cell fragments, of the selected strains of species originally detected by screening.

The use of immobilised enzymes (see Chapter 3) is now familiar to industry and their advantages are well recognised, so the practicality of using the new enzymes in an immobilised form will be determined early in the screening procedure. If the enzyme is produced intracellularly, the feasibility of using it without isolation and purification will be considered very seriously and strains selected for their amenability to use in this way.

It should be emphasised that there will be a constant dialogue between laboratory scientists and biochemical process engineers from the earliest stages of the screening process. Once the biochemical engineers are satisfied that their initial criteria of productivity, activity and stability can be met, the selected strain(s) of microbe will be grown in pilot plant conditions. It is only by applying the type of equipment used in full-scale plants that accurate costing of processes can be achieved. Pilot studies will probably reveal imperfections, or at least areas of ignorance, that must be corrected at the laboratory scale. If this proves possible, the pilot plant will produce samples of the enzyme preparation to be used by customers, who may well also be at the pilot plant stage in the development of the enzyme-utilising process. The enzyme pilot plant also produces samples for safety and toxicological studies provided that the pilot process is exactly similar to the full-scale operation.

Screening for new enzymes is expensive so that the intellectual property generated must be protected against copying by competitors. This is usually done by patenting the enzyme or its production method or, most usefully, the process in which it is to be used. Patenting will be initiated as soon as there is evidence that an innovative discovery has been made.

Media for enzyme production

Detailed description of the development and use of fermenters for the large-scale cultivation of microorganisms for enzyme production is outside the scope of this volume but mention of media use is appropriate because this has a bearing on the cost of the enzyme and because media components often find their way into commercial enzyme preparations. Details of components used in industrial-scale fermentation broths for enzyme production are not readily obtained. This is not unexpected as manufacturers have no wish to reveal information that may be of technical or commercial value to their competitors. Also some components of media may be changed from batch to batch as availability and cost of, for instance, carbohydrate feedstocks change. Such changes reveal themselves in often quite profound differences in appearance from batch to batch of a single enzyme from a single producer. The effects of changing feedstocks must be considered in relation to downstream processing. If such variability is likely significantly to reduce the efficiency of the standard methodology, it may be economical to use a more expensive defined medium of easily reproducible composition.

Clearly defined media are usually out of the question for large-scale use on grounds of cost but may be perfectly acceptable when enzymes are to be produced for high-value uses, such as analysis or medical therapy, where very pure preparations are essential. Less-defined complex media are composed of ingredients selected on the basis of cost and availability as well as composition. Waste materials and by-products from the food and agricultural industries are often major ingredients. Thus molasses, corn steep liquor, distillers solubles and wheat bran are important components of fermentation media, providing carbohydrate, minerals, nitrogen and some vitamins. Extra carbohydrate is usually supplied as starch, sometimes refined but often simply as ground cereal grains. Soybean meal and ammonium salts are frequently used sources of additional nitrogen. Most of these materials will vary in quality and composition from batch to batch causing changes in enzyme productivity.

Preparation of enzymes

Readers of papers dealing with the preparation of enzymes for research purposes will be familiar with tables detailing the stages of purification. Often the enzyme may be purified several hundred-fold but the yield of the enzyme may be very poor, frequently below 10% of the activity of the original material (Table 2.2). In contrast, industrial enzymes will be purified

Table 2.2 *The effect of number of steps on the yield and costs in a typical enzyme purification process*

Step	Relative weight	Yield (%)	Specific activity	Total cost	Cost per weight	Cost per activity
	1.000	100	1	1.00	1	1.00
1	0.250	75	3	1.10	4	1.47
2	0.063	56	9	1.20	19	2.13
3	0.016	42	27	1.30	83	3.08
4	0.004	32	81	1.40	358	4.92
5	0.001	24	243	1.50	1536	6.32

The realistic assumptions are made that step yields are 75%, step purifications are three-fold and step costs are 10% of the initial costs (later purification steps are usually intrinsically more expensive but are necessarily of smaller scale).

as little as possible, only other enzymes and material likely to interfere with the process which the enzyme is to catalyse will be removed. Unnecessary purification will be avoided as each additional stage is costly in terms of equipment, manpower and loss of enzyme activity. As a result, some commercial enzyme preparations consist essentially of concentrated fermentation broth, plus additives to stabilise the enzyme's activity.

The content of the required enzyme should be as high as possible (e.g. 10% (w/w) of the protein) in order to ease the downstream processing task. This may be achieved by developing the fermentation conditions or, often more dramatically, by genetic engineering. It may well be economically viable to spend some time cloning extra copies of the required gene together with a powerful promoter back into the producing organism in order to obtain 'over-producers' (see Chapter 8).

It is important that the maximum activity is retained during the preparation of enzymes. Enzyme inactivation can be caused by heat, proteolysis, suboptimal pH, oxidation, denaturants, irreversible inhibitors and loss of cofactors or coenzymes. Of these, heat inactivation, together with associated pH effects, is probably the most significant. It is likely to occur during enzyme extraction and purification if insufficient cooling is available (see Chapter 1), but the problem is less when preparing thermophilic enzymes. Proteolysis is most likely to occur in the early stages of extraction and purification when the proteases responsible for protein turnover in living cells are still present. It is also the major reason for enzyme inactivation by microbial contamination. In their native conformations, enzymes have highly structured domains which are resistant to attack by proteases because

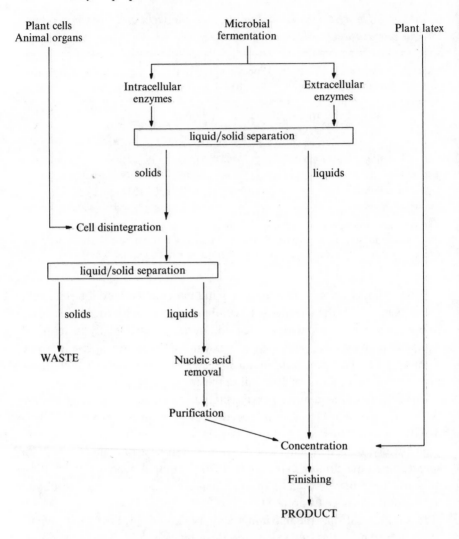

Figure 2.1. Flow diagram for the preparation of enzymes.

many of the peptide bonds are mechanically inaccessible and because many proteases are highly specific. The chances of a susceptible peptide bond in a structured domain being available for protease attack are low. Single 'nicks' by proteases in these circumstances may have little immediate effect on protein conformation and, therefore, activity. The effect, however, may severely reduce the conformational stability of the enzyme to heat or pH variation so greatly reducing its operational stability. If the domain is unfolded under these changed conditions, the whole polypeptide chain may be available for proteolysis and the same, specific, protease may destroy it. Clearly the best way of preventing proteolysis is rapidly to remove, or to inhibit, protease activity. Before this can be achieved it is important to keep enzyme preparations cold to maintain the native conformation of the molecules and slow any protease action that may occur.

Some intracellular enzymes are used commercially without isolation and purification but the majority of commercial enzymes are either produced extracellularly by the microbe or plant or must be released from the cells into solution and processed further (Figure 2.1). Solid/liquid separation is generally required for the initial separation of cell mass, the removal of cell debris after cell breakage and the collection of precipitates. This can be achieved by filtration, centrifugation or aqueous biphasic partition. In general, filtration or aqueous biphasic systems are used to remove unwanted cells or cell debris whereas centrifugation is the preferred method for the collection of required solid material.

Centrifugation

Centrifugation separates on the basis of the particle size and density difference between the liquid and solid phases. Sedimentation of material in a centrifugal field may be described by:

$$v = \frac{d^2(\rho_s - \rho_l)\omega^2 r F_s}{18\eta\theta} \tag{2.1}$$

where v is the rate of sedimentation, d is the particle diameter, ρ_s is the particle density, ρ_l is the solution density, ω is the angular velocity in radians s^{-1}, r is the radius of rotation, η is the kinematic viscosity, F_s is a correction factor for particle interaction during hindered settling and θ is a shape factor ($= 1$ for spherical particles). F_s depends on the volume fraction of the solids present; approximately equalling 1, 0.5, 0.1 and 0.05 for 1%, 3%, 12% and 20% solids volume fraction, respectively. Only material which reaches a surface during the flow through continuous centrifuges will be removed from the centrifuge feedstock, the efficiency depending on the residence time within the centrifuge and the distance necessary for sedimentation (D). This residence time will equal the volumetric throughout (ϕ) divided by the volume of the centrifuge (V). The maximum throughput of a centrifuge for efficient use is given by:

$$\phi = \frac{d^2(\rho_s - \rho_l)\omega^2 r V F_s}{18\eta\theta D} \tag{2.2}$$

The efficiency of the process is seen to depend on the solids volume fraction, the effective clarifying surface (V/D) and the acceleration factor

(a)

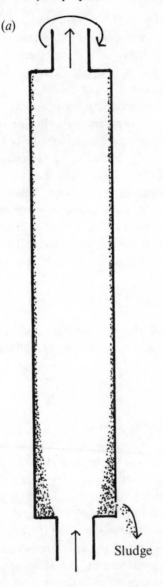

Sludge

Figure 2.2 Basic designs of industrial centrifuges, showing the flow of material within the bowls. Motor drives, cooling jackets and sludge collection vessels are not shown. (a) Tubular bowl centrifuge. This is generally operated vertically, the tubular rotor providing a long flow path, enabling clarification. The sludge collects and must be removed. (b) Continuous scroll centrifuge. This is operated horizontally. The helical screw scrolls the solids along the bowl surface and out of the liquid, the sludge being dewatered before discharge. The clarified liquor overflows

(*b*)

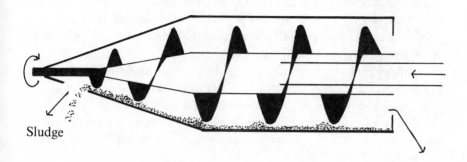

Sludge

(*c*)

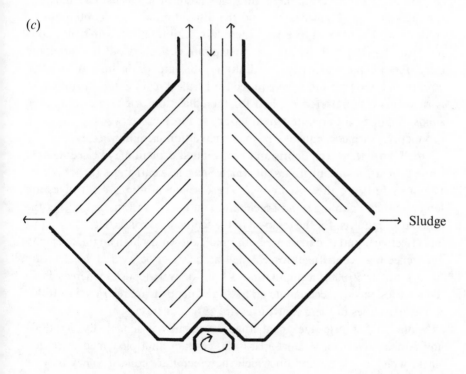

Sludge

Figure 2.2 (*cont.*)
over an adjustable weir at the other end of the bowl. The screw conveyor
rotates at a speed slightly different from that of the bowl. (*c*) Continuous
multichamber disc-stack centrifuge. The bowl contains a number of
parallel discs providing a large clarifying surface with a small
sedimentation distance. The sludge is removed through a valve.

($\omega^2 r/g$, where g is the gravitational constant; a rotor of radius 25 cm spinning at 1 rev. s^{-1} has an acceleration factor of approximately 1). Low acceleration factors of about 1500 may be used for harvesting cells whereas much higher acceleration factors are needed to collect enzyme efficiently. The product of these factors ($\omega^2 r V/gD$) is called the sigma factor (Σ) and is used to compare centrifuges and to assist scale-up.

Laboratory centrifuges using tubes in swing-out or angle head rotors have high angular velocity (ω) and radius of rotation (r) but small capacity (V) and substantial sedimentation distance (D). This type of design cannot be scaled-up safely, primarily because the mechanical stress on the centrifuge head increases with the square of the radius, which must increase with increasing capacity

For large-scale use, continuous centrifuges of various types are employed (Figure 2.2). These allow the continuous addition of feedstock, the continuous removal of supernatant and the discontinuous, semicontinuous or continuous removal of solids. Where discontinuous or semicontinuous removal of precipitate occurs, the precipitate is flushed out by automatic discharge systems, which cause its dilution with water or medium and may be a problem if the precipitate is required for further treatment. Centrifugation is the generally preferred method for the collection of enzyme-containing solids as it does not present a great hazard to most enzymes so long as foam production, with consequent enzymic inactivation, is minimised.

Small particles of cell debris and precipitated protein may be sedimented using tubular bowl centrifuges, of which Sharples centrifuges (produced by Pennwalt Ltd) are the best known. These semicontinuous centrifuges are long and thin, enabling rapid acceleration and deceleration, minimising the down-time required for the removal of the sedimented solids. Here the radius and effective liquid thickness are both small, allowing a high angular velocity and hence high centrifugal force; small models can be used at acceleration factors up to 50000, accumulating 0.1 kg of wet deposit, whereas large models, designed to accumulate up to 5 kg of deposit, are restricted to 16000 g. The capacities of these centrifuges are only moderate.

Multichamber disc-stack centrifuges, originally designed (by Westfalia and Alpha-Laval) for cream separation, contain multiple coned discs in a stack, which are spun and on which the precipitate collects. They may be operated either semi-continuously or, by using a centripetal pressurising pump within the centrifuge bowl, which forces the sludge out through a valve continuously. The capacity and radius of such devices are large and the thickness of liquid is very small, due to the large effective surface area. The angular velocity, however, is restricted giving a maximum acceleration factor of about 8000. A different design, which is rather similar in principle, is the solid bowl scroll centrifuge in which an Archimedes' screw collects the precipitate so that fluid and solids leave at opposite ends of the apparatus. These can only be used at low acceleration (about 3000 g) so they are suitable only for the collection of comparatively large particles.

Although many types of centrifuge are available, the efficient precipitation of small particles of cell debris can be difficult, sometimes nearly impossible. Clearly from equation (2.2) the efficiency of centrifugation can be improved if the particle diameter (d) is increased. This can be done either by coagulating or flocculating particles. Coagulation is caused by the removal of electrostatic charges (e.g. by pH change) and allowing particles to adhere to each other. Flocculation is achieved by adding small amounts of high molecular weight charged materials which bridge oppositely charged particles to produce a loose aggregate which may be readily removed by centrifugation or filtration. Flocculation and coagulation are cheap and effective aids to precipitating or otherwise harvesting whole cells, cell debris or soluble proteins but, of course, it is essential that the agents used must not inhibit the target enzymes. It is important to note that the choice of flocculant is determined by the pH and ionic strength of the solution and the nature of the particles. Most flocculants have very definite optimum concentrations above which further addition may be counter-effective. Some flocculants can be ruined rapidly by shear.

A comparatively recent introduction designed for the removal of cell debris is a moderately hydrophobic product in which cellulose is lightly derivatised with diethylaminoethyl functional groups. This material (Whatman CDR; cell debris remover) is inexpensive (essential as it is not reusable), binds to unwanted negatively charged cell constituents, acts as a filter aid and may be incinerated to dispose of hazardous wastes.

Filtration

Filtration separates simply on the basis of particle size. Its efficiency is limited by the shape and compressibility of the particles, the viscosity of the liquid phase and the maximum allowable pressures. Large-scale simple filtration employs filter cloths and filter aids in a plate and frame press configuration, in rotary vacuum filters (Figure 2.3) or centrifugal filters. The volumetric throughout of a filter is proportional to the pressure (P) and filter area (A_F) and inversely proportional to the filter cake thickness (D_F) and the dynamic viscosity:

$$\phi = \frac{kPA_F}{\eta D_F} \tag{2.3}$$

Spray

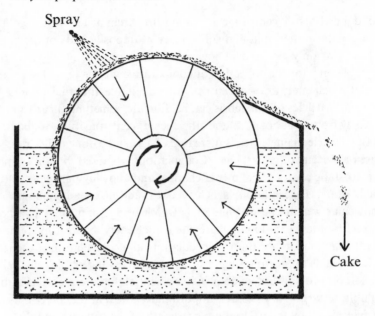

Cake

Figure 2.3. The basic design of the rotary vacuum filter. The suspension is
sucked through a filter cloth on a rotating drum. This produces a filter
cake, which is removed with a blade. The filter cake may be rinsed during
its rotation. These filters are generally rather messy and difficult to
contain, making them generally unsuitable for use in the production of
toxic or recombinant DNA products.

where k is a proportionality constant dependent on the size and nature of the
particles. For very small particles, k depends on the fourth power of their
diameter. Filtration of particles that are easily compressed leads to filter
blockage and the failure of equation (2.3) to describe the system. Under these
circumstances a filter aid, such as celite, is mixed with the feedstock to
improve the mechanical stability of the filter cake. Filter aids are generally
used only where the liquid phase is required as they cause substantial
problems in the recovery of solids. They also may cause loss of enzyme
activity from the solution due to physical hold-up in the filter cake. It is often
difficult for a process development manager to decide whether to attempt to
recover enzyme trapped in this way. Problems associated with the build-up of
the filter cake may also be avoided by high tangential flow of the feedstock
across the surface of the filter, a process known as *crossflow microfiltration*
(Figure 2.4). This method dispenses with filter aids and uses special symme-
tric microporous membrane assemblies capable of retaining particles down
to 0.1–1 µm diameter (cf. *Bacillus* diameter of about 2 µm).

(*a*)   (*b*)

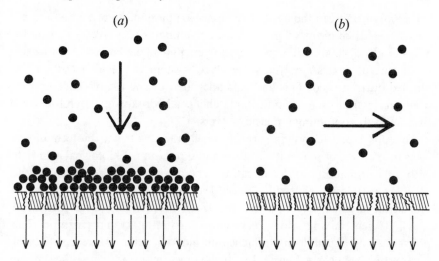

Figure 2.4. Principles of (*a*) dead-end filtration and (*b*) crossflow filtration. In dead-end filtration the flow causes the build-up of the filter cake, which may prevent efficient operation. This is avoided in cross-flow filtration where the flow sweeps the membrane surface clean.

A simple and familiar filtration apparatus is the perforated bowl centrifuge or basket centrifuge, in effect a spin drier. Cell debris is collected on a cloth with, or without, filter aid and can be skimmed off when necessary using a suitable blade. Such centrifugal filters have a large radius and effective liquid depth, allowing high volumes. However, safety decrees that the angular velocity must be low and so only large particles (e.g. plant material) can be removed satisfactorily.

Aqueous biphasic systems

The 'incompatibility' of certain polymers in aqueous solution was first noted by Beijerinck in 1896. In this case two phases were formed when agar was mixed with soluble starch or gelatine. Since then, many two-phase aqueous systems have been found, the most thoroughly investigated being the aqueous dextran–polyethylene glycol system (e.g. 10% (w/v) polyethylene glycol 4000/2% (w/v) dextran T500), where dextran forms the more hydrophilic, denser, lower phase and polyethylene glycol the more hydrophobic, less dense, upper phase. Aqueous three-phase systems are also known.

Phases form when limiting concentrations of the polymers are exceeded. Both phases contain mainly water and are enriched in one of the polymers. The limiting concentrations depend on the type and molecular weight of the polymers and on the pH, ionic strength and temperature of the solution.

Some polymers form the upper hydrophobic phase in the presence of fairly concentrated solutions of phosphates or sulphates (e.g. 10% (w/v) polyethylene glycol 4000/12.5% (w/v) potassium phosphate buffer). A drawback to the useful dextran–polyethylene glycol system is the high cost of the purified dextran used. This has been alleviated by the use of crude unfractionated dextran preparations, much cheaper hydroxpropyl starch derivatives and salt-containing biphasic systems.

Aqueous biphasic systems are of considerable value to biotechnology. They provide the opportunity for the rapid separation of biological materials with little probability of denaturation. The interfacial tension between the phases is very low (i.e. about 400-fold less than that between water and an immiscible organic solvent), allowing small droplet size, large interfacial areas, efficient mixing under very gentle stirring and rapid partition. The polymers have a stabilising influence on most proteins. A great variety of separations has been achieved, by far the most important being the separation of enzymes from broken crude cell material. Separation may be achieved in a few minutes, minimising the harmful action of endogenous proteases. The systems have also been used successfully for the separation of different types of cell membranes and organelles, the purification of enzymes and for extractive biconversions (see Chapter 7). Continuous liquid two-phase separation is easier than continuous solid/liquid separation using equipment familiar from immiscible solvent systems, for example disc-stack centrifuges and counter-current separators. Such systems are readily amenable to scale-up and may be employed in continuous enzyme extraction processes involving some recycling of the phases.

Cells, cell debris, proteins and other material distribute themselves between the two phases in a manner described by the partition coefficient (P) defined as:

$$P = \frac{C_t}{C_b} \tag{2.4}$$

where C_t and C_b represent the concentrations in the top and bottom phases, respectively. The yield and efficiency of the separation is determined by the relative amounts of material in the two phases and therefore depends on the volume ratio (V_t/V_b). The partition coefficient is exponentially related to the surface area (and hence molecular weight) and surface charge of the particles, in addition to the difference in the electrical potential and hydrophobicity of the phases. It is not generally very sensitive to temperature changes. This means that proteins and larger particles are normally partitioned into one phase, whereas smaller molecules are distributed more evenly between phases. A partition coefficient of greater than 3 is required if usable yields are to be achieved by a single extraction process. Typical partition coefficients for proteins are 0.01–100, whereas the partition coefficients for cells and cell debris are effectively zero.

The influence of pH and salts on protein partition is complex, particularly when phosphate buffers are present. A given protein distributes differently between the phases at different pH values and ionic strengths but the presence of phosphate ions affects the partition coefficient in an anomalous fashion because these ions distribute themselves unequally, resulting in electrostatic potential (and pH) differences. This means that systems may be 'tuned' to enrich an enzyme in one phase, ideally the upper phase, with cell debris and unwanted enzymes in the lower phase.

An enzyme may be extracted from the upper (polyethylene glycol) phase by the addition of salts or further polymer, generating a new biphasic system. This stage may be used to purify the enzyme further. A powerful modification of this technique is to combine phase partitioning and affinity partitioning. Affinity ligands (e.g. triazine dyes) may be coupled to either polymer in an aqueous biphasic system and thus greatly increase the specificity of the extraction.

Cell breakage

Various intracellular enzymes are used in significant quantities and must be released from cells and purified (Table 2.1). The amount of energy that must be put into the breakage of cells depends very much on the type, and to some extent on the physiology of the organism. Some types of cell are broken readily by gentle treatment such as osmotic shock (e.g. animal cells and some Gram-negative bacteria such as *Azotobacter* species), whilst others are highly resistant to breakage. The latter include yeasts, green algae, fungal mycelia and some Gram-positive bacteria which have cell wall and membrane structures capable of resisting internal osmotic pressure of approx. 2 MPa (20 atm) and therefore have the strength, weight for weight, of reinforced concrete. Consequently a variety of cell disruption techniques have been developed involving solid or liquid shear or cell lysis.

The rate of protein released by mechanical cell disruption is usually found to be proportional to the amount of releasable protein:

$$\frac{dP}{dt} = -kP \tag{2.5}$$

where P is the protein content remaining associated with the cells, t is the time and k is a release rate constant dependent on the system. Integrating from $P = P_m$ (maximum protein releasable) at time zero to $P = P_t$ at time t gives:

$$\int_{P_m}^{P_t} \frac{dP}{P} = \int_0^t -k\,dt \tag{2.6}$$

Table 2.3   *Hazards likely to damage enzymes during cell disruption*

Heat	All mechanical methods require a large input of energy, generating heat. Cooling is essential for most enzymes. The presence of substrates, substrate analogues or polyols may also help to stabilise the enzyme
Shear	Shear forces are needed to disrupt cells and may damage enzymes, particularly in the presence of heavy-metal ions and/or an air interface
Proteases	Disruption of cells will inevitably release degradative enzymes which may cause serious loss of enzyme activity. Such action may be minimised by increased speed of processing, with as much cooling as possible. This may be improved by the presence of an excess of alternative substrates (e.g. inexpensive protein) or inhibitors in the extraction medium
pH	Buffered solutions may be necessary. The presence of substrates, substrate analogues or polyols may also help to stabilise the enzyme
Chemical	Some enzymes may suffer conformational changes in the presence of detergent and/or solvents. Polyphenolics derived from plants are potent inhibitors of enzymes. This problem may be overcome by the use of adsorbents, such as polyvinylpyrrolidone, and by the use of ascorbic acid to reduce polyphenol oxidase action
Oxidation	Reducing agents (e.g. ascorbic acid, mercaptoethanol and dithiothreitol) may be necessary
Foaming	The gas–liquid phase interfaces present in foams may disrupt enzyme conformation
Heavy-metal toxicity	Heavy-metal ions (e.g. iron, copper and nickel) may be introduced by leaching from the homogenisation apparatus. Enzymes may be protected from irreversible inactivation by the use of chelating reagents such as EDTA

therefore:

$$\ln(P_m/P_t) = kt \tag{2.7}$$

As the protein released from the cells (P_r) is given by:

$$P_r = P_m - P_t \tag{2.8}$$

the following equation for cell breakage is obtained:

$$\ln\left(\frac{P_m}{P_m - P_r}\right) = kt \tag{2.9}$$

It is most important in choosing cell disruption strategies to avoid damaging the enzymes. The particular hazards to enzyme activity relevant to cell breakage are summarised in Table 2.3. The most significant of these, in general, are heating and shear.

Media for enzyme extraction will be selected on the basis of cost-effectiveness, and so will include as few components as possible. Media will usually be buffered at a pH value which has been determined to give the maximum stability of the enzyme to be extracted. Other components will combat other hazards to the enzyme, primarily factors causing denaturation (Table 2.3).

Ultrasonic cell disruption

The treatment of microbial cells in suspension with inaudible ultrasound (greater than about 18 kHz) results in their inactivation and disruption. Ultrasonication utilises the rapid sinusoidal movement of a probe within the liquid. It is characterised by high frequency (18 kHz–1MHz), small displacements (less than about 50 μm), moderate velocities (a few m s^{-1}) and very high acceleration (up to about 80 000 g). Ultrasonication produces cavitation phenomena when acoustic power inputs are sufficiently high to allow the multiple production of microbubbles at nucleation sites in the fluid. The bubbles grow during the rarefying phase of the sound wave, then are collapsed during the compression phase. On collapse, a violent shock wave passes through the medium. The whole process of gas bubble nucleation, growth and collapse due to the action of intense sound waves is called cavitation. The collapse of the bubbles converts sonic energy into mechanical energy in the form of shock waves equivalent to several thousand atmospheres (300 MPa) pressure. This energy imparts motions to parts of cells which disintegrate when their kinetic energy content exceeds the wall strength. An additional factor which increases cell breakage is the microstreaming (very high velocity gradients causing shear stress) which occur near radially vibrating bubbles of gas caused by the ultrasound.

Much of the energy absorbed by cell suspensions is converted to heat so effective cooling is essential. The amount of protein released by sonication has been shown to follow equation (2.9). The constant (k) is independent of cell concentrations up to high levels and is approximately proportional to the input acoustic power above the threshold power necessary for cavitation. Disintegration is independent of the sonication frequency except insofar as the cavitation threshold frequency depends on the frequency.

Equipment for the large-scale continuous use of ultrasonics has been available for many years and is widely used by the chemical industry but has not yet found extensive use in enzyme production. Reasons for this may be the conformational lability of some (perhaps most) enzymes to sonication and the damage that they may realise though oxidation by the free radicals, singlet oxygen and hydrogen peroxide that may be produced concomitantly.

Use of radical scavengers (e.g. N_2O) have been shown to reduce this inactivation. As with most cell breakage methods, very fine cell debris particles may be produced which can hinder further processing. Sonication remains, however, a popular, useful and simple small-scale method for cell disruption.

High-pressure homogenisers

Various types of high-pressure homogeniser are available for use in the food and chemicals industries but the design which has been very extensively used for cell disruption is the Manton–Gaulin APV type homogeniser. This consists of a positive displacement pump which draws cell suspension (about 12% (w/v)) through a check valve into the pump cylinder and forces it, at high pressures of up to 150 MPa (10 tons per square inch) and flow rates of up to 10000 $1 \, h^{-1}$, through an adjustable discharge valve which has a restricted orifice (Figure 2.5). Cells are subjected to impact, shear and a severe pressure drop across the valve but the precise mechanism of cell disruption is not clear. The main disruptive factor is the pressure applied and consequent pressure drop across the valve. This causes the impact and shear stress, which are proportional to the operating pressure.

As narrow orifices which are vulnerable to blockage are key parts of this type of homogeniser, it is unsuitable for the disruption of mycelial organisms but has been used extensively for the disruption of unicellular organisms. The release of proteins can be described by equation (2.9) but normally a similar relationship is used where the time variable is replaced by the number of passes (N) through the homogeniser:

$$\ln\left(\frac{P_m}{P_m - P_r}\right) = kN \tag{2.10}$$

In the commonly used operating range with pressures below about 75 MPa, the release constant (k) has been found to be proportional to the pressure raised to an exponent dependent on the organism and its growth history (e.g. $k = k'P^{2.9}$ in *Saccharomyces cerevisiae* and $k = k'P^{2.2}$ in *Escherichia coli*, where P is the operating pressure and k' is a rate constant). Different growth media may be selected to give rise to cells of different cell wall strength. Clearly, the higher the operating pressure, the more efficient is the disruption process. The protein release rate constant (k) is temperature dependent, disruption being more rapid at higher temperatures. In practice, this advantage cannot be used, since the temperature rise due to adiabatic compression is very significant, so samples must be pre-cooled and cooled again between multiple passes. At an operating pressure of 50 MPa, the temperature rise at each pass is about 12 deg. C.

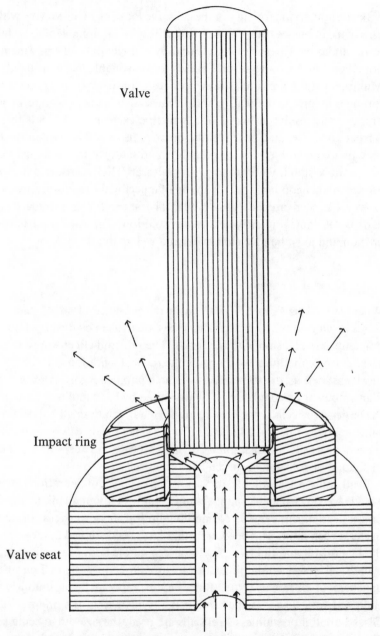

Figure 2.5. A cross-section through the Manton–Gaulin homogeniser valve, showing the flow of material. The cell suspension is pumped at high pressure through the valve, impinging on it and the impact ring. The shape of the exit nozzle from the valve seat varies between models and appears to be a critical determinant of the homogenisation efficiency. The model depicted is the 'CD Valve' from APV Gaulin.

In addition to the fragility of the cells, the location of an enzyme within the cells can influence the conditions of use of a homogeniser. Unbound intracellular enzymes may be released by a single pass whereas membrane-bound enzymes require several passes for reasonable yields to be obtained. Multiple passes are undesirable because, of course, they decrease the throughput productivity rate and because the further passage of already broken cells results in fine debris, which is excessively difficult to remove further downstream. Consequently, homogenisers will be used at the highest pressures compatible with the reliability and safety of the equipment and the temperature stability of the enzyme(s) released. High-pressure homogenisers are acceptably good for the disruption of unicellular organisms, provided the enzymes needed are not heat labile. The shear forces produced are not capable of damaging enzymes free in solution. The valve unit is prone to erosion and must be precision made and well maintained.

Use of bead mills

When cell suspensions are agitated in the presence of small steel or glass beads (usually 0.2–1.0 mm diameter) they are broken by the high liquid shear gradients and collision with the beads. The rate and effectiveness of enzyme release can be modified by changing the rates of agitation and the size of the beads, as well as the dimensions of the equipment. Any type of biomass, filamentous or unicellular, may be disrupted by bead milling but, in general, the larger-sized cells will be broken more readily than small bacteria. For the same volume of beads, a large number of small beads will be more effective than a relatively small number of larger beads because of the increased likelihood of collisions between beads and cells.

Bead mills are available in various sizes and configurations, from the Mickle shaker, which has a maximum volume of about 40 ml, to continuous process equipment capable of handling up to 200 kg wet yeast or 20 kg wet bacteria each hour. The bead mills that have been studied in most detail are the Dyno-Mill and the Netsch-Molinex agitator, both of which consist of a cylindrical vessel containing a motor-driven central shaft equipped with impellers of different types. Both can be operated continuously, being equipped with devices which retain the beads within the milling chamber. Glass Ballotini or stainless steel balls are used, the size range being selected for most effective release of the enzyme required. Thus 1-mm diameter beads are satisfactory for the rapid release of periplasmic enzymes from yeast but 0.25-mm diameter beads must be used, for a longer period, to release membrane-bound enzymes from bacteria.

The kinetics of protein release from bead mills follows the relation given by equation (2.9) with respect to the time (t) that a particle spends in the mill. Unfortunately, however well designed these mills are, when continuously operated there will be a significant amount of backmixing, which reduces the efficiency of the protein released with respect to the average residence time (τ, see the discussion concerning backmixing in reactors in Chapter 5). This is more noticeable at low flow rates (high average residence times) and when the proportion of protein released is high. It may be counteracted by designing the bead mill to encourage plug flow characteristics. Under these circumstances the relationship can be shown to be:

$$\ln\left(\frac{P_m}{P_m - P_r}\right) = \frac{\ln(1 + ik\tau)}{i} \tag{2.11}$$

where i represents the degree of backmixing (i.e. $i = 0$ under ideal plug flow conditions and $i = 1$ for ideal complete backmixing). Equation (2.11) reduces to give the simplified relationship of equation (2.9) at low (near zero) values of i.

In addition to bead size, the protein release rate constant (k) is a function of temperature, bead loading, impeller rotational speed and cell loading. Impeller speeds can be increased with advantage until bead breakage becomes significant but heat generation will also increase. At a constant impeller speed, the efficiency of the equipment declines with throughput as the degree of backmixing increases. There will be an optimum impeller tip speed at which the increases in disruption are balanced by increases in backmixing.

In general, increased bead loading increases the rate of protein release but also increases the production of heat and the power consumption. Heat production is the major problem in the use of bead mills for enzyme release, particularly on a large (e.g. 20 litres) scale. Smaller vessels may be cooled adequately through cooling jackets around the bead chamber, but larger mills require cooling through the agitator shaft and impellers. However, if cooling is effective there is little damage to the enzymes released.

Use of freeze-presses

The Hughes press and the 'X' press enables frozen cell pastes to be forced under high pressure (150–230 MPa (10–15 tons per square inch)) through narrow orifices, the disruption being produced by phase and volume changes and by solid shear due to the ice crystals. The Hughes press can only be used discontinuously on a small scale. The 'X' press may be used semicontinuously and is amenable to scale-up. However, although the method allows the breakage of even the most robust organisms and the efficient recovery of heat-sensitive enzymes, freeze-pressing is not used on a large scale for releasing enzymes from cells.

Use of lytic methods

The breakage of cells using non-mechanical methods is attractive in that it offers the prospects of releasing enzymes under conditions that are gentle, do not subject the enzyme to heat or shear, may be very cheap, and are quiet to the user. The methods that are available include osmotic shock, freezing followed by thawing, cold shock, desiccation, enzymic lysis and chemical lysis. Each method has its drawbacks but may be particularly useful under certain specific circumstances.

Certain types of cell can be caused to lyse by osmotic shock. This would be a cheap, gentle and convenient method of releasing enzymes but has not apparently been used on a large scale. Some types of cell may be caused to autolyse, in particular yeasts and *Bacillus* species. Yeast invertase preparations employed in the industrial manufacture of invert sugars are produced in this manner. Autolysis is a slow process compared with mechanical methods, and microbial contamination is a potential hazard, but it can be used on a very large scale if necessary. Where applicable, desiccation may be very useful in the preparation of enzymes on a large scale. The rate of drying is very important in these cases, slow methods being preferred to rapid ones, such as lyophilisation.

Enzymic lysis using added enzymes has been used widely on the laboratory scale but is less popular for industrial purposes. Lysozyme, from hen egg-white, is the only lytic enzyme available on a commercial scale. It has often used to lyse Gram-positive bacteria in an hour at about 50000 U kg^{-1} (dry weight). The chief objection to its use on a large scale is its cost. Where costs are reduced by the use of the relatively inexpensive, lysozyme-rich, dried egg-white, a major separation problem may be introduced. Yeast-lytic enzymes from *Cytophaga* species have been studied in some detail and other lytic enzymes are under development. If significant markets for lytic enzymes are identified, the scale of their production will increase and their cost is likely to decrease.

Lysis by acid, alkali, surfactants and solvents can be effective in releasing enzymes, provided that the enzymes are sufficiently robust. Detergents, such as Triton X-100, used alone or in combination with certain chaotropic agents, such as guanidine·HCl, are effective in releasing membrane-bound enzymes. However, such materials are costly and may be difficult to remove from the final product.

Preparation of enzymes from clarified solution

In many cases, especially when extracellular enzymes are being prepared for sale, the clarified solution is simply concentrated, preservative materials added, and the resulting product sold as a solution or as a dried preparation. The concentration process chosen will be the cheapest that is compatible with the retention of enzyme activity. For some enzymes, rotary evaporation can be considered, followed if necessary by spray drying. The most popular method, though, is *ultrafiltration*, whereby water and low molecular weight materials are removed by passage through a membrane under pressure, enzyme being retained. Ultrafiltration differs from conventional filtration and microfiltration with respect to the size of particles being retained (< 50 nm diameter). It uses asymmetric microporous membranes with a relatively dense but thin skin, containing pores, supported by a coarse strong substructure. Membranes possessing molecular weight cut-offs from 1000 to 100000 and usable at pressures up to 2 MPa are available.

There are a number of types of apparatus available. Stirred cells represent the simplest configuration of ultrafiltration cell. The membrane rests on a rigid support at the base of a cylindrical vessel, which is equipped with a magnetic stirrer to combat concentration polarisation. It is not suitable for large-scale use but is useful for preliminary studies and for the concentration of laboratory column eluates. Various large-scale units are available in which membranes are formed into wide diameter tubes (1–2 cm diameter) and the tubes grouped into cartridges. These are not as compact as capillary systems (area/volume about 25 m^{-1}) and are very expensive but less liable to blockage by stray large particles in the feedstream. Cheaper thin-channel systems are available (area/volume about 500 m^{-1}) which use flat membrane sandwiches in filter press arrangements of various designs chosen to produce laminar flow across the membrane and minimise concentration polarisation. Capillary membranes represent a relatively cheap and increasingly popular type of ultrafiltration system which uses microtubular membranes 0.2–1.1 mm diameter and provides large membrane areas within a small unit volume (area/volume about 1000 m^{-1}). Membranes are usually mounted into modules for convenient manipulation. This configuration of membranes can be scaled-up with ease. Commercial models are available that give ultrafiltration rates of up to 600 1 h^{-1}.

The steady improvement in the performance, durability and reliability of membranes has been a boon to enzyme technologists, encouraging wide use of the various ultrafilitration configurations. Problems with membrane blockage and fouling can usually be overcome by treatment of membranes with detergents, proteases or, with care, acids or alkalis. The initial cost of membranes remains considerable but modern membranes are durable and cost-effective. Ultrafiltration, done efficiently, results in little loss of enzyme activity. However, some configurations of apparatus, particularly in which solutions are recycled, can produce sufficient shear to damage some enzymes.

Nucleic acid removal

Intracellular enzyme preparations contain nucleic acids, which can give rise to increased viscosity, interfering with enzyme purification procedures, in particular ultrafiltration. Some organisms contain sufficient nuclease activity to eliminate this problem, but otherwise the nucleic acids must be removed by precipitation or degraded by the addition of exogenous nucleases. Ammonium sulphate precipitation (see later) can be effective in removing nucleic acids but will remove some protein at the same time. Various more specific precipitants have been used, usually positively charged materials which form complexes with the negatively charged phosphate residues of the nucleic acids. These include, in order of roughly decreasing effectiveness, polyethyleneimine, the cationic detergent cetyltrimethyl ammonium bromide, streptomycin sulphate and protamine sulphate. All of these are expensive and possibly toxic, particularly streptomycin sulphate. Also, they may complex undesirably with certain enzymes. They may be necessary, however, where possible contamination of the enzyme product must be avoided, such as in the preparation of restriction endonucleases. Otherwise, treatment with bovine pancreatic nucleases is probably the most cost-effective method of nucleic acid removal.

Concentration by precipitation

Precipitation of enzymes is a useful method of concentration and is ideal as an initial step in their purification. It can be used on a large scale and is less affected by the presence of interfering materials than are any of the chromatographic methods described later.

Salting-out of proteins particularly by use of ammonium sulphate, is one of the best known and used methods of purifying and concentrating enzymes, particularly on the laboratory scale. Increases in the ionic strength of the solution cause a reduction in the repulsive effect of like charges between identical molecules of a protein. It also reduces the forces holding the solvation shell around the protein molecules. When these forces are sufficiently reduced, the protein will precipitate; hydrophobic proteins precipitating at lower salt concentrations than hydrophilic proteins. Ammonium sulphate is convenient and effective because of its high solubility, cheapness, lack of toxicity to most enzymes and its stabilizing effect on some enzymes (see Table 2.4). Its large-scale use, however, is limited, as it is corrosive except with stainless steel, it forms dense solutions presenting problems to the collection of the precipitate by centrifugation, and it may release gaseous ammonia, particularly at alkaline pH. The practice of using ammonium sulphate precipitation is more straightforward than the theory. Reproducible results can only be obtained provided the protein concentration, temperature and pH are kept constant. The concentration of the salt needed to precipitate an enzyme will vary with the concentration of the enzyme. However, fractionation of protein mixtures by a stepwise increase in the ionic strength can be a very effective way of partly purifying enzymes.

The solubility of an enzyme can be described by:

$$\log S = K_{\text{intercept}} - K_{\text{salt}}T \tag{2.12}$$

where S is the enzyme solubility, $K_{\text{intercept}}$ is the intercept constant, K_{salt} is the salting out constant and T is the ionic strength, which is proportional to the concentration of a precipitating salt. $K_{\text{intercept}}$ is independent of the salt used but depends on the pH, temperature, enzyme and the other components in the solution. K_{salt} depends on both the enzyme required and the salt used but is largely independent of other factors. Equation (2.12) may also be used to give the minimum salt concentration necessary before an enzyme will start to precipitate, the concentration change necessary to precipitate the enzyme varying according to the magnitude of the salting-out constant.

Some enzymes do not survive ammonium sulphate precipitation. Other salts may be substituted but the more favoured alternative is to use organic solvents such as methanol, ethanol, propan-2-ol and acetone. These act by reducing the dielectric constant of the medium and consequently reducing the solubility of proteins by favouring protein–protein rather than protein–solvent interactions. Organic solvents are not widely used on a large scale because of their cost, their inflammability, and the tendency of proteins to undergo rapid denaturation by these solvents if the temperature is allowed to rise much above 0 °C. On safety grounds, when organic solvents are used, special flameproof laboratory areas are used and temperatures maintained below the solvents' flashpoints.

Except when enzymes are presented for sale as ammonium sulphate precipitates, the precipitating salt or solvent must be removed. This may be done by dialysis, ultrafiltration or by using a desalting column of, for instance, Sephadex G–25.

Heat treatment

In many cases, unwanted enzyme activities may be˙ removed by heat treatment. Different enzymes have differing susceptibility to heat denaturation and precipitation. Where the enzyme required is relatively heat stable, this allows its easy and rapid purification in terms of enzymic activity. For such enzymes heat treatment is always considered as an option at an early stage in their purification. This method has been particularly successfully applied to the production of glucose isomerase, where a short incubation at a relatively high temperature is used (e.g. 60–85 °C for 10 min). No interfering activity remains after this treatment and the heat-treated, and hence leaky, cells may be immobilised and used directly.

Chromatography

Enzyme preparations that have been clarified and concentrated are now in a suitable state for further purification by chromatography. For enzyme purification there are three principal types of chromatography utilising the ion-exchange, affinity and gel exclusion properties of the enzyme, usually in that order. Ion-exchange and affinity chromatographic methods can both rapidly handle large quantities of crude enzyme but ion-exchange materials are generally cheaper and, therefore, preferred at an earlier stage in the purification where the scale of operation is somewhat greater. Gel exclusion chromatography (also sometimes called 'gel filtration' or just 'gel chromatography', although it does not separate by a filtering mechanism, larger molecules passing more rapidly through the matrix than smaller molecules) is relatively slow and has the least capacity and resolution. It is generally left until last as an important final purification step and also as a method of changing the solution buffer before concentration, finishing and sale. Where sufficient information has been gathered regarding the size and variation of charge with pH of the required enzyme and its major contaminants, a rational purification scheme can be devised. A relatively quick analytical method for obtaining such data utilises a two-dimensional process whereby electrophoresis occurs in one direction and a range of pH is produced in the other; movement in the electric field is determined by the size and sign of a protein's charge, which both depend on the pH. As the sample is applied across the range of pH, this method produces titration curves (i.e. charge versus pH) for all proteins present.

A large effort has been applied to the development of chromatographic matrices suitable for the separation of proteins. The main problem that has had to be overcome is that of ensuring that the matrix has a sufficiently large surface area available to molecules as large as proteins (i.e. they are macroporous) whilst remaining rigid and incompressible under rapid elution conditions. In addition, matrices must generally be hydrophilic and inert. Although the standard bead diameters of most of these matrices are non-uniform and fairly large (50–150 μm), many are now supplied as uniform-sized small beads (e.g. 4–6 μm diameter), which allows their use in very efficient separation processes (*high performance liquid chromatography*, HPLC), but at exponentially increasing cost with decreasing bead size. Relatively high pressures are needed to operate such columns, necessitating specialised equipment and considerable additional expense. They are used only for the small-scale production of expensive enzymes, where a high degree of purity is required (e.g. restriction endonucleases and therapeutic enzymes).

Column manufacturers now supply equipment for monitoring and controlling chromatography systems so that it is possible to have automated apparatus which loads the sample, collects fractions and regenerates the column. Such equipment must, of course, have fail-safe devices to protect both column and product.

Ion-exchange chromatography

Enzymes possess a net charge in solution, dependent upon the pH and their structure and isoelectric point. In solutions at a pH below their isoelectric point they will be positively charged and bind to cation exchangers, whereas in solutions at a pH above their isoelectric point they will be negatively charged and bind to anion exchangers. The pH chosen must be sufficient to maintain a high, but opposite, charge on both protein and ion-exchanger and the ionic strength must be sufficient to maintain the solubility of the protein without the salt being able to compete successfully with the protein for ion-exchange sites. The binding is predominantly reversible and its strength is determined by the pH and ionic strength of the solution and the structures of the enzyme and ion-exchanger. Normally the pH is kept constant and enzymes are eluted by increasing the solution ionic strength. A very wide range of ion-exchange resins, cellulose derivatives and large-pore gels are available for chromatographic use.

Ion-exchange materials are generally water-insoluble polymers containing cationic or anionic groups. Cation exchange matrices have anionic functional groups such as $-SO_3^-$, $-OPO_3^-$ and $-COO^-$ and anion exchange matrices usually contain the cationic tertiary and quaternary ammonium groups, with general formulae $-NHR_2^+$ and $-NR_3^+$. Proteins become bound by exchange with the associated counter-ions.

Ion-exchange polystyrene resins are eminently suitable for large-scale chromatographic use but have low capacities for proteins owing to their small pore size. Binding is often strong, due to the resin hydrophobicity, and the conditions needed to elute proteins are generally severe and may be denaturing. Nevertheless such resins are a potential means of concentrating or purifying enzymes.

Ion-exchange cellulose and large pore gels are more generally suitable for enzyme purification and, indeed, many were designed for that task. A variety of charged groups, anionic or cationic, may be introduced. The practical level of substitution of cellulose is limited, as derivatisation above 1 mol kg^{-1} may lead to dissolution of the cellulose. Consequently, proteins may be eluted from them under mild conditions. Ion-exchange cellulose can be used in both batch and column processes, but on a large scale they are used mainly batchwise (see below). This is because the increased speed of large-scale batchwise processing and the avoidance of the deep-bed filtering characteristics of columns outweigh any advantage due to the increase in resolution on columns. Careful preparation before use, and regeneration after use are essential for their effective use.

Batchwise operations involve stirring the pretreated and equilibriated ion-exchanger with the enzyme solution in a suitably cooled vessel. Adsorption to the exchangers is usually rapid (e.g. less than 30 min) but some proteins can take far longer to adsorb completely. Stirring is essential but care must be taken not to generate fine particles (fines). Unadsorbed material may be removed in a variety of manners. Basket centrifuges are a particularly convenient means of hastening the removal of the initial supernatant and the elution of the adsorbed material. This is usually done using stepwise increases in ionic strength and/or changes in pH but it is possible to place the exchangers, plus adsorbed material, in a column and elute using a suitable gradient. However, whilst ion-exchange cellulose are widely used for column chromatography on the laboratory scale, their compressibility causes difficulty when attempts are made to use large-scale columns.

Some of the problems with derivatised cellulose may be overcome using more recently introduced materials. Derivatives of cross-linked agarose (Sepharose CL-6B) and of the synthetic polymer Trisacryl have high capacities (up to 150 mg protein ml^{-1}), yet are not significantly compressible. In addition, they do not change volume with pH and ionic strength, which allows them to be regenerated without removal from the chromatographic column.

Affinity chromatography

This is a term which now covers a variety of methods of enzyme purification, the common factor of which is the more or less specific interaction between the enzyme and the immobilised ligand. In its most specific form, the immobilised ligand is a substrate or competitive inhibitor of the enzyme. Ideally it should be possible to purify an enzyme from a complex mixture in a single step and, indeed, purification factors of up to several thousand-fold have been achieved. An alternative, equally specific, approach is to use an antibody to the enzyme as the ligand. Such specific matrices, though, are very expensive and cannot be employed generally on a large scale. Additionally, they often do not perform as well as might be expected due to non-specific binding effects. In general, affinity chromatography achieves a higher purification factor (with a median value in reported purifications of about ten-fold) than ion-exchange chromatography (with a median performance of about three-fold), in spite of it generally being used at a later stage in the purification when there is less purification possible.

A less specific approach, suitable for many enzymes, is to use analogues of coenzymes, such as NAD^+, as the ligand. This method has been used successfully but has now been superceded by the employment of a series of water-soluble dyes as ligands. These are much cheaper and, usually by trial and error, have been found to have surprising degrees of specificity for a wide range of enzymes. This dye-affinity chromatography was allegedly discovered by accident, certain enzymes being found to bind to the blue-dyed dextran used as a molecular weight standard to calibrate gel exclusion columns.

Another fortuitous discovery was hydrophobic interaction chromatography, found when it was noted that certain proteins were unexpectedly retained on affinity columns containing hydrophobic spacer arms. Hydrophobic adsorbents now available include octyl or phenyl groups. Hydrophobic interactions are strong at high solution ionic strength so samples need not be desalted before application to the adsorbent. Elution is achieved by changing the pH or ionic strength or by modifying the dielectric constant of the eluent, using, for instance, ethanediol. A recent introduction is cellulose, derivatised to introduce even more hydroxyl groups. This material (Whatman HB1) is designed to interact with proteins by hydrogen bonding. Samples are applied to the matrix in a concentrated (over 50% saturated, > 2 M) solution of ammonium sulphate. Proteins are eluted by diluting the ammonium sulphate. This introduces more water which competes with protein for the hydrogen bonding sites. The selectivity of both of these methods is similar to that of fractional precipitation using ammonium sulphate but their resolution may be somewhat improved by their use in chromatographic columns rather than batchwise.

Careful choice of matrices for affinity chromatography is necessary. Particles should retain good flow and porosity properties after attachment of the ligands and should not be capable of the non-specific adsorption of proteins. Agarose beads fulfil these criteria and are readily available as ligand supports (see also Chapter 3). Affinity chromatography is not used extensively in the large-scale manufacture of enzymes, primarily because of cost. Doubtless as the relative costs of materials are lowered, and experience in handling these materials is gained, enzyme manufacturers will make increased use of these very powerful techniques.

Gel exclusion chromatography

There is now a considerable choice of materials that can separate proteins on the basis of their molecular size. The original cross-linked dextrans (Sephadex G- series, Pharmacia Ltd) and polyacrylamides (Bio-Gel P- series, BioRad Ltd) are still, quite rightly, widely used. Both types are available in a wide range of pore sizes and particle size distribution. However, as the pore size increases, for use with larger enzymes, these gels become progressively less rigid and therefore less suitable for large-scale use. Consequently, alternative, but generally more costly, rigid gel materials have been developed for the fractionation of proteins of molecular weight greater than about 75000. These are the cross-linked derivatives of agarose (Sepharose CL and Superose) and dextran (Sephacryl S) made by Pharmacia Ltd, the cross-linked polyacrylamide-agarose mixtures (Ultrogel AcA) made by LKB Instruments Ltd and the ethylene glycol-methacrylate copolymers (Fractogel HW) made by Toyo Soda Company (TSK). These are available in a range of forms capable of fractionating enzymes, and other materials, with molecular weights up to 10^8 and at high flow rates. Although these gels are described as 'rigid', it should be appreciated that this is a relative term. The best gels are significantly compressible so scale-up from laboratory-sized columns cannot be achieved by producing longer columns. Scale-up is achieved by increasing the diameter of columns (up to about 1 m diameter) but retaining the small depth. Further scale-up is done by connecting such sections in series to produce 'stacks'. Extreme care must be taken in packing all gel columns so as to allow even, well-distributed flow through the gel bed. For the same reason, the end pieces of the columns must allow even distribution of material over the whole surface of the column. The newer materials are supplied in a pre-swollen state which enable their rapid and efficient packing using slight pressure.

Gel exclusion chromatography invariably causes dilution of the enzyme, which must then be concentrated using one of the methods described earlier.

Preparation of enzymes for sale

Once the enzyme has been purified to the desired extent and concentrated, the manufacturer's main objective is to retain the activity. Enzymes for industrial use are sold on the basis of overall activity. Often a freshly supplied enzyme sample will have a higher activity than that stated by the manufacturer. This is done to ensure that the enzyme preparation has the guaranteed storage life. The manufacturer will usually recommend storage conditions and quote the expected rate of loss of activity under those conditions. It is of primary importance to the enzyme producer and customer that the enzymes retain their activity during storage and use. Some enzymes retain their activity under operational conditions for weeks or even months: most do not.

To achieve stability, the manufacturers use all the subtleties at their disposal. Formulation is an art and often the precise details of the methods used to stabilise enzyme preparations are kept secret or revealed to customers only under the cover of a confidentiality agreement. Sometimes it is only the formulation of an enzyme that gives a manufacturer the competitive edge over rival companies. It should be remembered that most industrial enzymes contain relatively little active enzyme (< 10% (w/w), including isoenzymes and associated enzyme activities), the rest being due to inactive protein, stabilisers, preservatives, salts and the diluent which allows standardisation between production batches of different specific activities.

The key to maintaining enzyme activity is maintenance of conformation, so preventing unfolding, aggregation and changes in the covalent structure. Three approaches are possible: (1) use of additives, (2) the controlled use of covalent modification, and (3) enzyme immobilisation (discussed further in Chapter 3).

In general, proteins are stabilised by increasing their concentration and the ionic strength of their environment. Neutral salts compete with proteins for water and bind to charged groups or dipoles. This may result in the inter-actions between an enzyme's hydrophobic areas being strengthened, causing the enzyme molecules to compress and making them more resistant to thermal unfolding reactions. Not all salts are equally effective in stabilising hydro-phobic interactions, some are much more effective at their destabilisation by binding to them and disrupting the localised structure of water (the *chaotropic effect*, Table 2.4). From this it can be seen why ammonium sulphate and potassium hydrogen phosphate are powerful enzyme stabilisers, whereas sodium thiosulphate and calcium chloride destabilise enzymes. Many enzymes are specifically stabilised by low concentrations of cations which may or may not form part of the active site: for example, Ca^{2+} stabilises α-amylases and Co^{2+} stabilises glucose isomerases. At high concentrations (e.g. 20% (w/v) NaCl), salt discourages microbial growth due to its osmotic effect. In addition ions can offer some protection against oxidation to groups such as thiols by salting-out the dissolved oxygen from solution.

Low molecular weight polyols (e.g. glycerol, sorbitol and mannitol) are

Table 2.4   *Effect of ions on enzyme stabilisation*

	Increased chaotropic effect ←	
Cations	Al^{3+}, Ca^{2+}, Mg^{2+}, Li^+, Na^+, K^+, NH_4^+, $(CH_3)_4N^+$	
Anions	SCN^-, I^-, ClO_4^-, Br^-, Cl^-, SO_4^{2-}, HPO_4^{2-}, $citrate^{3-}$	
	Increased stabilisation ———→	

also useful for stabilising enzymes, by repressing microbial growth, due to the reduction in the water activity, and by the formation of protective shells which prevent unfolding processes. Glycerol may be used to protect enzymes against denaturation due to ice-crystal formation at subzero temperatures. Some hydrophilic polymers (e.g. polyvinyl alcohol, polyvinylpyrrolidone and hydroxypropylcelluloses) stabilise enzymes by a process of compartmentation whereby the enzyme–enzyme and enzyme–water interactions are partly replaced by less potentially denaturing enzyme–polymer interactions. They may also act by stabilising the hydrophobic effect within the enzymes.

Many specific chemical modifications of amino acid side-chains are possible which may (or, more commonly, may not) result in stabilisation. A useful example of this is the derivatisation of lysine side-chains in proteases with *N*-carboxyamino acid anhydrides. These form polyaminoacylated enzymes with various degrees of substitution and length of amide-linked side-chains. This derivatisation is sufficient to disguise the proteinaceous nature of the protease and prevent autolysis.

Important lessons about the molecular basis of thermostability have been learned by comparison of enzymes from mesophilic and thermophilic organisms. A frequently found difference is the increase in the proportion of arginine residues at the expense of lysine and histidine residues. This may possibly be explained by noting that arginine is bidentate and has a higher pK_a than lysine or histidine (see Table 1.1, p. 13). Consequently, it forms stronger salt links with bidentate aspartate and glutamate side-chains, resulting in more rigid structures. This observation, among others, has given hope that site-specific mutagenesis may lead to enzymes with significantly improved stability (see Chapter 8). In the meantime it remains possible to convert lysine residues to arginine-like groups by reaction with activated ureas. It should be noted that enzymes stabilised by making them more rigid usually show lower activity (i.e. V_{max}) than the 'natural' enzyme.

Enzymes are very much more stable in the dry state than in solution. Solid enzyme preparations sometimes consist of freeze-dried protein. More usually they are bulked out with inert materials such as starch, lactose, carboxymethylcellulose and other polyelectrolytes which protect the enzyme during a cheaper spray-drying stage. Other materials which are added to enzymes before sale may be substrates, thiols to create a reducing environment, antibiotics, benzoic acid esters as preservatives for liquid enzyme preparations, inhibitors of contaminating enzyme activities and chelating agents. Additives of these types must, of course, be compatible with the final use of the enzyme's product.

Enzymes released on to the market should conform to a number of quality procedures including regulatory requirements, which are legal and mandatory. This is provided by the *quality assurance* (QA) within the company. Enzyme products must be consistent as appropriate to their intended use. This may be ensured by *good manufacturing practice* (GMP) and further checked by *quality control* (QC).

Customer service

A customer who is likely to use large quantities of an enzyme will be expected to specify the form and activity in which the enzyme is supplied. The development of a new enzyme-catalysed process is often a matter of teamwork between the customer and the enzyme company's development scientists. Once the process is running, the enzyme company will probably be in contact with the customer through three types of individual.

(1) Salespeople, who ensure that the supply of enzyme continues and that the cost of the enzyme is mutually acceptable.

(2) Technical salespeople, who liaise with technical managers to ensure that their product is performing up to specification. This association often leads to suggestions for the improvement of the process. The technical sales team will be expected by the customer to deal with problems to do with the enzyme. They may be able to solve the problems themselves or may require the services of the third group, the laboratory or pilot plant scientists.

(3) The laboratory and pilot plant scientists will spend some time troubleshooting as and when necessary. They may test materials for customers, for instance if a new source of raw material is under consideration.

All the enzyme company's employees must, of course, work as a team and it is in the interests of both customer and manufacturer if their technical experts also cooperate closely. Both organisations will learn from each other. Technical advances should accrue as a result of suggestions from both sides.

Safety and regulatory aspects of enzyme use

Only very few enzymes present hazards, because of their catalytic activity, to those handling them in normal circumstances, but there are several areas of potential hazard arising from their chemical nature and source. These are allergenicity, activity-related toxicity, residual microbiological activity and chemical toxicity.

All enzymes, being proteins, are potential allergens and have especially potent effects if inhaled as a dust. Once an individual has developed an immune response as a result of inhalation or skin contact with the enzyme, re-exposure produces increasingly severe responses becoming dangerous or even fatal. Because of this, dry enzyme preparations have been replaced to a large extent by liquid preparations, sometimes deliberately made viscous to lower the likelihood of aerosol formation during handling. Where dry preparations must be used, as in the formulation of many enzyme detergents, allergenic responses by factory workers are a very significant problem, particularly when fine dusting powders are employed. Workers in such environments are usually screened for allergies and respiratory problems. The problem has been largely overcome by encapsulating and granulating dry enzyme preparations, a procedure that has been applied most success-fully to the proteases and other enzymes used in detergents. Enzyme producers and users recognise that allergenicity will always be a potential problem and provide safety information concerning the handling of enzyme preparations. They stress that dust in the air should be avoided, so weighing and manipulation of dry powders should be carried out in closed systems. Any spilt enzyme powder should be removed immediately, after first mois-tening it with water. Any waste enzyme powder should be dissolved in water before disposal into the sewage system. Enzyme on the skin or inhaled should be washed with plenty of water. Liquid preparations are inherently safer, but it is important that any spilt enzyme is not allowed to dry, as dust formation can then occur. The formation of aerosols (e.g. by poor operating procedures in centrifugation) must be avoided as these are at least as harmful as powders.

Activity-related toxicity is much rarer but it must be remembered that proteases are potentially dangerous, particularly in concentrated forms and especially if inhaled. No enzyme has been found to be toxic, mutagenic or carcinogenic by itself, as might have been expected from its proteinaceous structure. However, enzyme preparations cannot be regarded as completely safe, as such dangerous materials may be present as contaminants, derived from the enzyme source or produced during its processing or storage.

The organisms used in the production of enzymes may themselves be sources of hazardous materials and have been the chief focus of attention by the regulatory authorities. In the USA, enzymes must be *Generally Regarded As Safe* (GRAS) by the FDA (Food and Drug Administration) in order to be used as a food ingredient. Such enzymes include α and β-amylase, bromelain, catalase, cellulase, ficin, α-galactosidase, glucoamylase, glucose isomerase, glucose oxidase, invertase, lactase, lipase, papain, pectinase, pepsin, rennet and trypsin. In the UK, the Food Additives and Contaminants Committee (FACC) of the Ministry of Agriculture, Fisheries and Food divided enzymes into five classes on the basis of their safety for presence in the foods and use in their manufacture.

Group A Substances that the available evidence suggests are acceptable for use in food.

Group B Substances that on the available evidence may be regarded as provisionally acceptable for use in food but about which further information must be made available within a specified time for review.

Group C Substances for which the available evidence suggests toxicity and which ought not to be permitted for use in food until adequate evidence of their safety has been provided to establish their acceptability.

Group D Substances for which the available information indicates definite or probable toxicity and which ought not to be permitted for use in food.

Group E Substances for which inadequate or no toxicological data are available and for which it is not possible to express an opinion as to their acceptability for use in food.

This classification takes into account the potential chemical toxicity from microbial secondary metabolites such as mycotoxins and aflotoxins. The growing body of knowledge on the long-term effects of exposure to these toxins is one of the major reasons for the tightening of the legislative controls.

The enzymes that fall into group A are exclusively plant and animal enzymes such as papain, catalase, lipase, rennet and various other proteases. Group B contains a very wide range of enzymes from microbial sources, many of which have been used in food or food processing for many hundreds of years. The Association of Microbial Food Enzyme Producers (AMFEP) has suggested subdivisions of the FACC's group B into:

Class a Microorganisms that have traditionally been used in food or in food processing, including *Bacillus subtilis, Aspergillus niger, Aspergillus oryzae, Rhizopus oryzae, Saccharomyces cerevisiae, Kluyveromyces fragilie, Kluyveromyces lactis* and *Mucor javanicus.*

Class b Microorganisms that are accepted as harmless contaminants present in food, including *Bacillus stearothermophilus, Bacillus licheniformis, Bacillus coagulans* and *Klebsiella aerogenes.*

Class c Microorganisms that are not included in classes a and b, including *Mucor miehei, Streptomyces albus, Trichoderma reesei, Actinoplanes missouriensis*, and *Penicillium emersonii.*

It was proposed that class *a* should not be subjected to testing and that classes *b* and *c* should be subjected to the following tests: (1) acute oral toxicity in mice and rats, (2) subacute oral toxicity for four weeks in rats, (3) oral toxicity for three months in rats, and (4) *in vitro* mutagenicity. In addition class *c* should be tested for microorganism pathogenicity and, under exceptional circumstances, *in vivo* mutagenicity, teratogenicity, and carcinogenicity.

The cost of the various tests needed to satisfy the legal requirements are very significant and must be considered during the determination of process costs. Plainly the introduction of an enzyme from a totally new source will be a very expensive matter. It may prove more satisfactory to clone such an enzyme into one of AMFEP's class *a* organisms but this will first require new legislation to regulate the use of cloned microbes in foodstuffs. Some of the safety problems associated with the use of free enzymes may be overcome by using immobilised enzymes (see Chapter 3). This is an extremely safe technique, so long as the materials used are acceptable and neither they, nor the immobilised enzymes, leak into the product stream.

The production of enzymes is subject, in the UK, to the Health and Safety at Work Act 1974, to ensure the health and safety of employees. Good manufacturing practice is employed and controls ensure that enzyme production is performed by a pure culture of the producing microbes.

Summary

(*a*) Enzymes may be prepared from many sources but most are obtained by the fermentation of microorganisms.

(*b*) The industrial use of enzymes depends on their effectiveness, cost and safety.

(*c*) The rate of enzyme released by any mechanical homogenisation of cells is normally proportional to the amount of enzyme available.

(*d*) Centrifugation is generally used for the collection of solid enzymic material, whereas filtration is preferred for liquid enzyme recovery. Aqueous biphasic systems are becoming more commonly encountered in enzyme recovery operations.

(*e*) Preparation of industrial enzymes involves the minimum number of purification stages that is compatible with their use. Extensive purification is a very expensive process. Care must be taken over the prevention of inactivation during enzyme preparation.

(*f*) Enzymes offered for sale must be as stable as may be necessary and safe to handle.

Bibliography

Anon. (1986). A strategy for protein purification. *Separation News*, vol. 13.6, pp. 1–6. Uppsala: Pharmacia.

Atkinson, T., Scawen, M. D. & Hammond, P. M. (1987). Large scale industrial techniques of enzyme recovery. In *Biotechnology*, vol. 7a *Enzyme technology*, ed. J. F. Kennedy, pp. 279–323. Weinheim: VCH Verlagsgesellschaft mbH.

Barker, S. A. (1982). New approaches to enzyme stabilisation. In *Topics in enzyme and fermentation biotechnology*, vol. 6, ed. A. Wiseman, pp. 68–78. Chichester: Ellis Horwood Ltd.

Bonnerjea, J., Oh, S., Hoare, M. & Dunnill, P. (1986). Protein purification: the right step at the right time. *Biotechnology*. 4, 954–958.

Booth, A. G. (1987). *Protein purification: a strategic approach*. Oxford: IRL Press Ltd. [This is a computer-aided learning package suitable for IBM computers.]

Cejka, A. (1985). Preparation of media. In *Biotechnology*, vol. 2 *Fundamentals of biochemical engineering*, ed. H. Brauer, pp. 629–98. Weinheim: VCH Verlagsgesellschaft mbH.

Cheetham, P. S. J. (1987). Screening for novel biocatalysts. *Enzyme and Microbial Technology*, 9, 194–213.

Chisti, Y. & Moo-Young, M. (1987). Disruption of microbial cells for intracellular products. *Enzyme and Microbial Technology*, 8, 194–204.

Kula, M.-R. (1985). Recovery operations. In *Biotechnology*, vol. 2 *Fundamentals of biochemical engineering*, ed. H. Brauer, pp. 725–60. Weinheim: VCH Verlagsgesellschaft mbH.

Noordervliet, P.F. & Toet, D. A. (1987). Safety in enzyme technology. In *Biotechnology*, vol. 7a *Enzyme technology*, ed. J. F. Kennedy, pp. 711–41. Weinheim: VCH Verlagsgesellschaft mbH.

Strathmann, H. (1985). Membranes and membrane processes in biotechnology. *Trends in Biotechnology*, 3, 112–18.

3 The preparation and kinetics of immobilised enzymes

The economic argument for immobilisation

An important factor determining the use of enzymes in a technological process is their expense. Several hundred enzymes are commercially available at prices of about £1 mg^{-1}, although some are much cheaper and many are much more expensive. As enzymes are catalytic molecules, they are not used up directly by the processes in which they are involved. Their high initial cost, therefore, should only be incidental to their use. However, due to denaturation, they do lose activity with time. If possible, they should be destabilised against denaturation and utilised in an efficient manner. When they are used in a soluble form, they retain some activity after the reaction, which cannot be economically recovered for re-use and is generally wasted. This activity residue remains to contaminate the product and its removal may involve extra purification costs. In order to eliminate this wastage, and give an improved productivity, simple and economic methods must be used which enable the enzyme to be separated from the reaction product. The easiest way of achieving this is by separating the enzyme and product during the reaction using a two-phase system: one phase containing the enzyme and the other containing the product. The enzyme is imprisoned within its phase, allowing its re-use or continuous use but preventing it from contaminating the product; other molecules, including the reactants, are able to move freely between the two phases. This is best known as *immobilisation* and may be achieved by fixing the enzyme to, or within, some other material. The term immobilisation does not necessarily mean that the enzyme cannot move freely within its particular phase, although this is often the case. A wide variety of insoluble materials, also known as substrates (not to be confused with the enzymes' reactants), may be used to immobilise the enzymes by making them insoluble. These are usually inert polymeric or inorganic matrices.

Immobilisation of enzymes often incurs an additional expense and is only undertaken if there is a sound economic or process advantage in the use of the immobilised, rather than free (soluble), enzymes. The most important benefit derived from immobilisation is the easy separation of the enzyme from the products of the catalysed reaction. This prevents the enzyme from contaminating the product, minimising downstream processing costs and possible effluent-handling problems, particularly if the enzyme is noticeably toxic or antigenic. It also allows continuous processes to be practicable, with a considerable saving in enzyme, labour and overhead costs. Immobilisation often affects the stability and activity of the enzyme, but conditions are usually available where these properties are little changed or even enhanced. The productivity of an enzyme, so immobilised, is greatly increased, as it may be more fully used at higher substrate concentrations for longer periods than is the free enzyme. Insoluble immobilised enzymes are of little use, however, where any of the reactants are also insoluble, due to steric difficulties.

Methods of immobilisation

There are four principal methods available for immobilising enzymes: (1) adsorption, (2) covalent binding, (3) entrapment, and (4) membrane confinement (Figure 3.1).

Carrier matrices for enzyme immobilisation by adsorption and covalent binding must be chosen with care. Of particular relevance to their use in industrial processes is their cost relative to the overall process costs; ideally they should be cheap enough to discard. The manufacture of high-value products on a small scale may allow the use of relatively expensive supports and immobilisation techniques, whereas these would not be economical in the large-scale production of low added-value materials. A substantial saving in costs occurs where the carrier may be regenerated after the useful lifetime of the immobilised enzyme. The surface density of binding sites, together with the volumetric surface area sterically available to the enzyme, determines the maximum binding capacity. The actual capacity will be affected by the number of potential coupling sites in the enzyme molecules and the electrostatic charge distribution and surface polarity (i.e. the hydrophobic–hydrophilic balance) on both the enzyme and support. The nature of the support will also have a considerable affect on an enzyme's expressed activity and apparent kinetics. The form, shape, density, porosity, pore size distribution, operational stability and particle size distribution of the supporting matrix will influence the reactor configuration in which the immobilised biocatalyst may be used. The ideal support is cheap, inert, physically strong and stable. It will increase the enzyme specificity (k_{cat}/K_m) whilst reducing product inhibition, shift the pH optimum to the desired value for the process, and discourage microbial growth and non-specific adsorption. Some matrices possess other properties which are useful for particular purposes

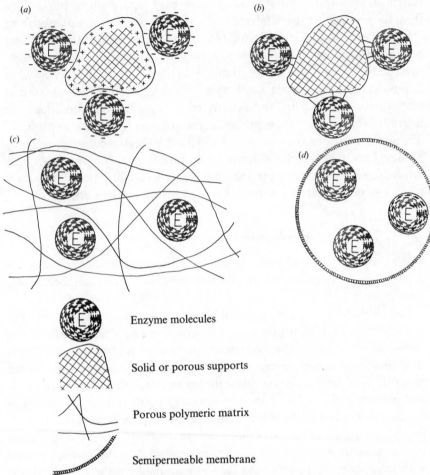

Figure 3.1. Immobilised-enzyme systems. (*a*) Enzyme non-covalently adsorbed to an insoluble particle; (*b*) enzyme covalently attached to an insoluble particle; (*c*) enzyme entrapped within an insoluble particle by a cross-linked polymer; (*d*) enzyme confined within a semipermeable membrane.

such as ferromagnetism (e.g. magnetic iron oxide, enabling transfer of the biocatalyst by means of magnetic fields), a catalytic surface (e.g. manganese dioxide, which catalytically removes the inactivating hydrogen peroxide produced by most oxidases), or a reductive surface environment (e.g. titania, for enzymes inactivated by oxidation). Clearly most supports possess only some of these features, but a thorough understanding of the properties of immobilised enzymes does allow suitable engineering of the system to approach these optimal qualities.

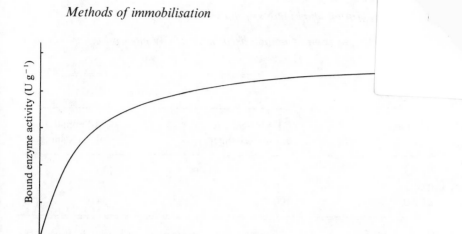

Figure 3.2. Diagram showing the effect of soluble enzyme concentration on the activity of enzyme immobilised by adsorption to a suitable matrix. The amount adsorbed depends on the incubation time, pH, ionic strength, surface area, porosity, and the physical characteristics of both the enzyme and the support.

Adsorption of enzymes on to insoluble supports is a very simple method of wide applicability, capable of high enzyme loading (about 1 g g^{-1} matrix). Simply mixing the enzyme with a suitable adsorbent, under appropriate conditions of pH and ionic strength, followed, after a sufficient incubation period, by washing off loosely bound and unbound enzyme will produce the immobilised enzyme in a directly usable form (Figure 3.2). The driving force causing this binding is usually due to a combination of hydrophobic effects and the formation of several salt links per enzyme molecule. The particular choice of adsorbent depends principally upon minimising leakage of the enzyme during use. Although the physical links between the enzyme molecules and the support are often very strong, they may be reduced by many factors including the introduction of the substrate. Care must be taken that the binding forces are not weakened during use, e.g. by inappropriate changes in pH or ionic strength. Examples of suitable adsorbents are ion-exchange matrices (Table 3.1), porous carbon, clays, hydrous metal oxides, glasses and polymeric aromatic resins. Ion-exchange matrices, although more expensive than these other supports, may be used economically due to the ease with which they may be regenerated when their bound enzyme has come to the end of its active life, a process which may simply involve washing off the used enzyme with concentrated salt solutions and re-suspending the ion-exchanger in a solution of active enzyme.

Table 3.1   *Preparation of immobilised invertase by adsorption (Woodward, 1985)*

% bound at	Support type	
	DEAE-Sephadex anion exchanger	CM-Sephadex cation exchanger
pH 2.5	0	100
pH 4.7	100	75
pH 7.0	100	34

The preparations are made by mixing 400 U of commercial invertase with 50 mg of the support in 10 ml of buffer (T < 20 mM) for 30 min, followed by washing off any weakly adsorbed enzyme. The specific activity of the adsorbed enzyme is about 40% of the original preparation.

Immobilisation of enzymes by their covalent coupling to insoluble matrices is an extensively researched technique. Only small amounts of enzyme may be immobilised by this method (about 0.02 g g^{-1} matrix) although in exceptional cases as much as 0.3 g g^{-1} matrix has been reported. The strength of binding is very strong, however, and very little leakage of enzyme from the support occurs. The relative usefulness of various groups found in enzymes, for covalent link formation, depends upon their availability and reactivity (nucleophilicity), in addition to the stability of the covalent link, once formed (Table 3.2). The reactivity of the protein side-chain nucleophiles is determined by their state of protonation (i.e. charged status) and roughly follows the relationship:

$$-S^- > -SH > -O^- > -NH_2 > -COO^- > -OH \gg -NH_3{}^+$$

where the charges may be estimated from a knowledge of the pK_a values of the ionising groups (Table 1.1, p. 13) and the pH of the solution. Lysine residues are found to be the most generally useful groups for covalent bonding of enzymes to insoluble supports, due to their widespread surface exposure and high reactivity, especially in slightly alkaline solutions. They also appear to be only very rarely involved in the active sites of enzymes.

The most commonly used method for immobilising enzymes on the research scale (i.e. using < 1 g of enzyme) involves Sepharose, activated by cyanogen bromide. This is a simple, mild and often successful method of wide applicability. Sepharose is a commercially available beaded polymer which is highly hydrophilic and generally inert to microbiological attack. Chemically it is an agarose (poly($\beta - 1,3$-D-galactose-$\alpha - 1,4 - (3,6 -$ anhydro)-L-galactose)) gel. The hydroxyl groups of this polysaccharide combine with

Table 3.2 *Relative usefulness of enzyme residues for covalent coupling*

Residue	Content	Exposure	Reactivity	Stability of couple	Use
Aspartate	+	+ +	+	+	+
Arginine	+	+ +	−	±	−
Cysteine	−	±	+ +	−	−
Cystine	+	−	±	±	−
Glutamate	+	+ +	+	+	+
Histidine	±	+ +	+	+	+
Lysine	+ +	+ +	+ +	+ +	+ +
Methionine	−	−	±	−	−
Serine	+ +	+	±	+	±
Threonine	+ +	±	±	+	±
Tryptophan	−	−	−	±	−
Tyrosine	+	−	+	+	+
C terminus	−	+ +	+	+	+
N terminus	−	+ +	+ +	+ +	+
Carbohydrate	− ~ + +	+ +	+	+	±
Others	− ~ + +	−	−	− ~ + +	−

+ + , + , ± and − denote high to low.

cyanogen bromide to give the reactive cyclic imido-carbonate. This reacts with primary amino groups (i.e mainly lysine residues) on the enzyme under mildly basic conditions (pH 9–11.5, Figure 3.3(a)). The high toxicity of cyanogen bromide has led to the commercial, if rather expensive, production of ready-activated Sepharose and the investigation of alternative methods, often involving chloroformates, to produce similar intermediates (Figure 3.3(b)). Carbodiimides (Figure 3.3(c)) are very useful bifunctional reagents as they allow the coupling of amines to carboxylic acids. Careful control of the reaction conditions and choice of carbodiimide allow a great degree of selectivity in this reaction. Glutaraldehyde is another bifunctional reagent which may be used to cross-link enzymes or link them to supports (Figure 3.3(d)). It is particularly useful for producing immobilised enzyme membranes, for use in biosensors, by cross-linking the enzyme plus a non-catalytic diluent protein within a porous sheet (e.g. lens tissue paper or nylon net fabric). The use of trialkoxysilanes allows even such apparently inert materials as glass to be coupled to enzymes (Figure 3.3(e)). There are numerous other methods available for the covalent attachment of enzymes (e.g. the attachment of tyrosine groups through diazo-linkages, and lysine groups through amide formation with acyl chlorides or anhydrides).

It is clearly important that the immobilised enzyme retains as much

(*a*)

cyanogen bromide
CNBr
Sepharose
very reactive cyanate ester
inert carbamate
reactive cyclic imido-carbamate
H₂N – enzyme
H₂N – enzyme
N-substituted carbamate
isourea derivative

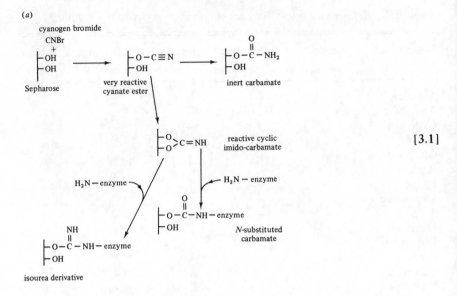

[3.1]

(*b*)

ethyl chloroformate
ClCO₂C₂H₅
H₂N – enzyme
cellulose
cyclic carbonate intermediate

[3.2]

(*c*)

carbodiimide
R – N=C=N – R′
H₂N – enzyme
matrix
O-acyl isourea
inert

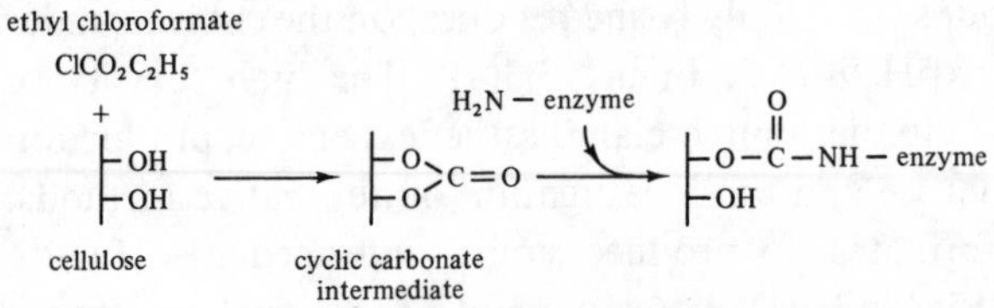

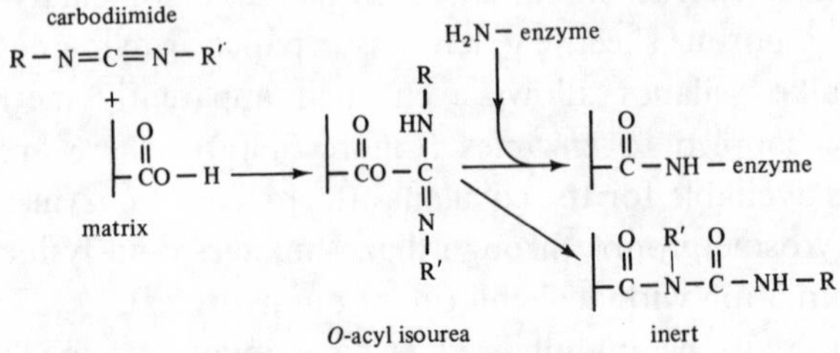

[3.3]

(d)

glutaraldehyde   oligoglutaraldehyde

$$
\begin{array}{l}
CH=O \\
\ | \\
CH_2 \\
\ | \\
CH_2 \\
\ | \\
CH_2 \\
\ | \\
CH=O
\end{array}
\ \rightleftharpoons \
\begin{array}{c}
\quad\quad CH=O \quad\quad\quad CH=O \quad\quad\quad\quad CH=O \\
\quad\quad\ | \quad\quad\quad\quad\quad\ | \quad\quad\quad\quad\quad\quad\ | \\
-CH-C-CH_2-CH_2-CH=C-CH_2-CH_2-CH=C-CH_2-CH_2-
\end{array}
$$

$$
\begin{array}{c}
\\
H_2N-\text{enzyme} \searrow \\
\\
\quad\quad CH=O \quad\quad\quad\quad CH=N-\text{enzyme} \quad\quad CH=O \\
\quad\quad\ | \quad\quad\quad\quad\quad\quad\quad\ | \quad\quad\quad\quad\quad\quad\ | \\
-CH-CH-CH_2-CH_2-CH=C-CH_2-CH_2-CH=C-CH_2-CH_2- \\
\quad\quad\ | \\
\quad\quad NH \\
\quad\quad\ | \\
\quad\text{enzyme}
\end{array}
$$

[3.4]

(e)

3-aminopropyltriethoxysilane

$$(C_2H_5O)_3Si(CH_2)_3NH_2$$

$$
\begin{array}{c}
\quad + \\
\quad\ | \\
\quad\ O \\
\quad\ | \\
-O-Si-OH \\
\quad\ | \\
\quad\ O \\
\quad\ | \\
-O-Si-OH \\
\quad\ | \\
\quad\text{glass}
\end{array}
\ \longrightarrow \
\begin{array}{c}
\ |\quad\quad\quad | \\
\ O\quad\quad\ O \\
\ |\quad\quad\quad | \\
-O-Si-O-Si-(CH_2)_3-NH_2 \\
\ |\quad\quad\quad | \\
\ O\quad\quad\ O \\
\ |\quad\quad\quad | \\
-O-Si-O-Si-(CH_2)_3-NH_2 \\
\ |\quad\quad\quad |
\end{array}
$$

thiophosgene ⟍
$$S=CCl_2$$

[3.5]

$$
\begin{array}{c}
\ |\quad\quad\quad | \\
\ O\quad\quad\ O \\
\ |\quad\quad\quad | \\
-O-Si-O-Si-(CH_2)_3-NCS \\
\ |\quad\quad\quad | \\
\ O\quad\quad\ O \\
\ |\quad\quad\quad | \\
-O-Si-O-Si-(CH_2)_3-NCS \\
\ |\quad\quad\quad |
\end{array}
$$

$$H_2N-\text{enzyme} \searrow$$

$$
\begin{array}{c}
\ |\quad\quad\quad |\quad\quad\quad\quad\quad S \\
\ O\quad\quad\ O\quad\quad\quad H\ \ || \\
\ |\quad\quad\quad |\quad\quad\quad\quad\ |\ \ | \\
-O-Si-O-Si-(CH_2)_3-N-C-NH-\text{enzyme} \\
\ |\quad\quad\quad |\quad\quad\quad\quad\quad S \\
\ O\quad\quad\ O\quad\quad\quad H\ \ || \\
\ |\quad\quad\quad |\quad\quad\quad\quad\ |\ \ | \\
-O-Si-O-Si-(CH_2)_3-N-C-NH-\text{enzyme} \\
\ |\quad\quad\quad |
\end{array}
$$

catalytic activity as possible after reaction. This can, in part, be ensured by reducing the amount of enzyme bound in non-catalytic conformations (Figure 3.4). Immobilisation of the enzyme in the presence of saturating concentrations of substrate, product or a competitive inhibitor ensures that the active site remains unreacted during the covalent coupling and reduces the occurrence of binding in unproductive conformations. The activity of the immobilised enzyme is then simply restored by washing the immobilised enzyme to remove these molecules.

Entrapment of enzymes within gels or fibres is a convenient method for use in processes involving low molecular weight substrates and products. Amounts in excess of 1 g enzyme g^{-1} gel or fibre may be entrapped. However, the difficulty which large molecules have in approaching the catalytic sites of entrapped enzymes precludes the use of entrapped enzymes with high molecular weight substrates. The entrapment process may be a purely physical caging or involve covalent binding. As an example of this latter method, the enzymes' surface lysine residues may be derivatised by reaction with acryloyl chloride ($CH_2\!=\!CH-CO-Cl$) to give the acryloyl amides. This product may then be co-polymerised and cross-linked with acrylamide   ($CH_2\!=\!CH-CO-NH_2$)   and   bisacrylamide ($H_2N-CO-CH\!=\!CH-CH\!=\!CH-CO-NH_2$) to form a gel. Enzymes may be entrapped in cellulose acetate fibres by, for example, making up an emulsion of the enzyme plus cellulose acetate in methylene chloride, followed by extrusion through a spinneret into a solution of an aqueous precipitant. Entrapment is the method of choice for the immobilisation of microbial, animal and plant cells where calcium alginate is widely used.

Membrane confinement of enzymes may be achieved by a number of quite different methods, all of which depend for their utility on the semipermeable

Figure 3.3. (See pp. 86–7.) Commonly used methods for the covalent immobilisation of enzymes (*a*) Activation of Sepharose by cyanogen bromide. Conditions are chosen to minimise the formation of the inert carbamate. (*b*) Chloroformates may be used to produce intermediates similar to those produced by cyanogen bromide but without its inherent toxicity. (*c*) Carbodiimides may be used to attach amino groups on the enzyme to carboxylate groups on the support or carboxylate groups on the enzyme to amino groups on the support. Conditions are chosen to minimise the formation of the inert substituted urea. (*d*) Glutaraldehyde is used to cross-link enzymes or link them to supports. It usually consists of an equilibrium mixture of monomer and oligomers. The product of the condensation of enzyme and glutaraldehyde may be stabilised against dissociation by reduction with sodium borohydride. (*e*) The use of trialkoxysilane to derivatise glass. The reactive glass may be linked to enzymes by a number of methods, including the use of thiophosgene as shown.

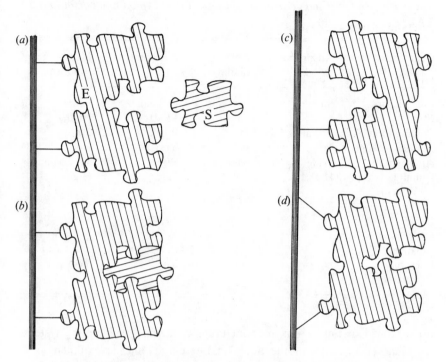

Figure 3.4. The effect of covalent coupling on the expressed activity of an immobilised enzyme. (*a*) Immobilised enzyme (E) with its active site unchanged and ready to accept the substrate molecule (S), as shown in (*b*). (*c*) Enzyme bound in a non-productive mode due to the inaccessibility of the active site. (*d*) Distortion of the active site produces an inactive immobilised enzyme. Non-productive modes are best prevented by the use of large molecules reversibly bound in or near the active site. Distortion can be prevented by use of molecules which can sit in the active site during the coupling process, or by the use of a freely reversible method for the coupling which encourages binding to the most energetically stable (i.e. native) form of the enzyme. Both (*c*) and (*d*) may be reduced by use of 'spacer' groups between the enzyme and support, effectively displacing the enzyme away from the steric influence of the surface.

nature of the membrane. This must confine the enzyme whilst allowing free passage for the reaction products and, in most configurations, the substrates. The simplest of these methods is achieved by placing the enzyme on one side of the semipermeable membrane whilst the reactant and product stream is present on the other side. Hollow-fibre membrane units are available commercially with large surface areas relative to their contained volumes (>20 m^2 l^{-1}) and permeable only to substances of molecular weight substantially less than the enzymes. Although costly, these are very easy to

Table 3.3   *Generalised comparison of different enzyme immobilisation techniques*

Characteristics	Adsorption	Covalent binding	Entrapment	Membrane confinement
Preparation	Simple	Difficult	Difficult	Simple
Cost	Low	High	Moderate	High
Binding force	Variable	Strong	Weak	Strong
Enzyme leakage	Yes	No	Yes	No
Applicability	Wide	Selective	Wide	Very wide
Running problems	High	Low	High	High
Matrix effects	Yes	Yes	Yes	No
Large diffusional barriers	No	No	Yes	Yes
Microbial protection	No	No	Yes	Yes

use for a wide variety of enzymes (including coenzyme-regenerating systems, see Chapter 8), without the additional research and development costs associated with other immobilisation methods. Enzymes encapsulated within small membrane-bound droplets or liposomes (see Chapter 7) may also be used within such reactors. As an example of the former, the enzyme is dissolved in an aqueous solution of 1,6-diaminohexane. This is then dispersed in a solution of hexanedioic acid in the immiscible solvent, chloroform. The resultant reaction forms a thin polymeric (Nylon–6,6) shell, around the aqueous droplets which traps the enzyme. Liposomes are concentric spheres of lipid membranes, surrounding the soluble enzyme. They are formed by the addition of phospholipid to enzyme solutions. The microcapsules and liposomes are washed free of non-confined enzyme and transferred back to aqueous solution before use.

Table 3.3 presents a comparison of the more important general characteristics of these methods.

Kinetics of immobilised enzymes

The kinetic behaviour of a bound enzyme can differ significantly from that of the same enzyme in free solution. The properties of an enzyme can be modified by suitable choice of the immobilisation protocol, whereas the same method may have appreciably different effects on different enzymes. These

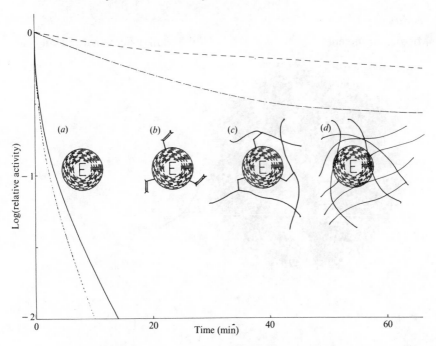

Figure 3.5. Illustration of the use of multipoint interactions for the stabilisation of enzymes (Martinek *et al*, 1977*a*, *b*). (*a*) ———— activity of free underivatised chymotrypsin. (*b*) ...... activity of chymotrypsin derivatised with acryloyl chloride. (*c*) ––––––– activity or acryloyl chymotrypsin copolymerised within a polymethacrylate gel. Up to 12 residues are covalently bound per enzyme molecule. Lower derivatisation leads to lower stabilisation. (*d*) –·––·––·– activity of chymotrypsin non-covalently entrapped within a polymethacrylate gel. The degree of stabilisation is determined by strength of the gel, and hence the number of non-covalent interactions. All reactions were performed at 60 °C using low molecular weight artificial substrates. The immobilised chymotrypsin preparations showed stabilisation of up to 100000-fold, most of which is due to their multipoint nature although the consequent prevention of autolytic loss of enzyme activity must be a significant contributory factor.

changes may be due either to conformational alterations within the enzyme, resulting from the immobilisation procedure, or to the presence and nature of the immobilisation support.

Immobilisation can greatly affect the stability of an enzyme. If the immobilisation process introduces any strain into the enzyme, this is likely to encourage the inactivation of the enzymes under denaturing conditions (e.g. higher temperatures or extremes of pH). However, where there is an unstrained multipoint binding between the enzyme and the support, substantial stabilisation may occur (Figure 3.5). This is due primarily to the

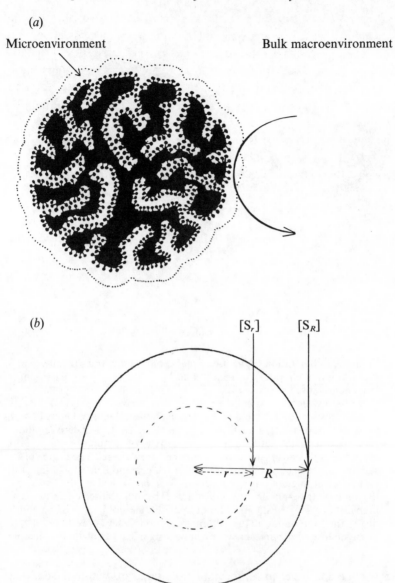

Figure 3.6. A schematic cross-section of an immobilised enzyme particle (*a*) shows the macroenvironment and microenvironment. ● represents the enzyme molecules. The microenvironment consists of the internal solution plus that part of the surrounding solution which is influenced by the surface characteristics of the immobilised enzyme. Partitioning of substances will occur between these two environments. Substrate molecules (S) must diffuse through the surrounding layer (external transport) in order to reach the catalytic surface and be converted to

physical prevention within the protein structure of the large conformational changes which generally precede its inactivation. Many successful covalent immobilisation processes involve an initial freely reversible stage, where the covalent links are allowed to form, break and re-form until an unstrained covalently linked structure is created, in order to stabilise the resultant immobilised enzyme. Additional stabilisation is derived by preventing the enzyme molecules from interacting with each other, and the protection that immobilisation affords towards proteolytic and microbiological attack. This later effect is due to a combination of diffusional difficulties and the camouflage to enzymic attack produced by the structural alterations. In order to achieve maximum stabilisation of the enzymes, the surfaces of the enzyme and support should be complementary, with the formation of many unstrained covalent or non-covalent interactions. Often, however, this factor must be balanced against others, such as the cost of the process, the need for a specific support material, and ensuring that the substrates are not sterically hindered from diffusing to the active site of the immobilised enzyme where they react at a suitable rate.

The kinetic constants (e.g. K_m, V_{max}) of enzymes may be altered by the process of immobilisation due to internal structural changes and restricted access to the active site. Thus, the intrinsic specificity (k_{cat}/K_m) of such enzymes may well be changed relative to the soluble enzyme. An example of this involves trypsin, where the freely soluble enzyme hydrolyses 15 peptide bonds in the protein pepsinogen but the immobilised enzyme hydrolyses only 10. The apparent value of these kinetic parameters, when determined experimentally, may differ from intrinsic values. This may be due to changes

Figure 3.6 (*cont.*)
product (P). In order for all the enzyme to be utilised, substrate must also diffuse within the pores in the surface of the immobilised-enzyme particle (internal transport). The porosity (ϵ) of the particle is the ratio of the volume of solution contained within the particle to the total volume of the particle. The tortuosity (τ) is the average ratio of the path length, via the pores, between any points within the particle to their absolute distance apart. The tortuosity, which is always greater than or equal to unity, clearly depends on the pore geometry. The diagram exaggerates dimensions for the purpose of clarity. Typically, the diameter of enzyme molecules (2–10 nm) are one to two orders of magnitude smaller than the pore diameters which are two to four orders of magnitude smaller than the particle diameters (10–2000 μm); the microenvironment consists of a diffusion layer (~ 10 μm thick) and a thinner partition layer (~ 20 nm thick). (*b*) The concentration of the substrate at the surface of the particles (radius R) is [S_R], whereas the internal concentration at any smaller radius (r) is the lower value represented by [S_r].

in the properties of the solution in the immediate vicinity of the immobilised enzyme, or to the effects of molecular diffusion within the local environment (Figure 3.6). The relationship between these parameters is shown below:

intrinsic parameters of the soluble enzyme
↓
intrinsic parameters of the immobilised enzyme
↓
apparent parameters due to the partition and diffusion

Effect of solute partition on the kinetics of immobilised enzymes

The solution lying within a few molecular diameters ($\approx$ 10 nm) from the surface of an immobilised enzyme will be influenced by both the charge and hydrophobicity of the surface. Charges are always present on the surface of immobilised enzyme particles due to the amphoteric nature of enzymes. Where these positive and negative charges are not equally balanced, the net charge on the surface exerts a considerable effect over the properties of the microenvironment. This surface charge, easily produced by the use of ion-exchange or similar charged matrices for the immobilisation, repels molecules of similar charge whilst attracting those possessing opposite charge. The force of attraction or repulsion due to this charge is significant over molecular distances but decays rapidly with the square of the distance from the surface. A partitioning of charged molecules (e.g. substrates and products) occurs between the bulk solution and the microenvironment; molecules of charge opposite to that of the immobilised enzyme surface are partitioned into the microenvironment, whereas molecules possessing charge similar to that of the immobilised enzyme surface are expelled, with equal effect, into the bulk solution. The solution partition may be quantified by introduction of the *electrostatic partition coefficient* (Λ) defined by:

$$\Lambda = \frac{[C_0^{n+}]}{[C^{n+}]} = \frac{[A^{n-}]}{[A_0^{n-}]}$$

(3.1)

where $[C_0^{n+}]$ and $[A_0^{n-}]$ represent each cation and anion bulk concentration, $[C^{n+}]$ and $[A^{n-}]$ represent their concentration within the microenvironment and n is the number of charges on each ion. Λ has been found to vary within the range about 0.01 to 100, Λ being greater than unity for positively charged enzymic surfaces and less than unity for negatively charged surfaces. The effect of partition on positively and negatively charged molecules is equal but opposite. For a positively charged support, cations are partitioned away from the microenvironment, whereas the concentration of anions is greater within this volume compared with that in the bathing solution. Λ depends on the density of charge on and within the immobilised enzyme particle. It is greatly influenced by the ionic strength of the solution. At high ionic strength the raised concentration of charged solute molecules counteracts the charge on the particles, reducing the electrostatic force and causing Λ to approach to unity. Assuming Michaelis–Menten kinetics, the rate of reaction catalysed by an immobilised enzyme is given by equation (1.9) (p. 9), where the substrate concentration is the concentration within the microenvironment:

$$v = \frac{V_{max}}{1 + K_m/[S]} \qquad (3.2)$$

If the substrate is positively charged, it follows from equation (3.1) that:

$$\Lambda = [S_o]/[S] \qquad (3.3)$$

therefore:

$$v = \frac{V_{max}}{1 + K_m^{app}/[S_o]} \qquad (3.4)$$

where:

$$K_m^{app} = K_m\Lambda \qquad (3.5)$$

K_m^{app} is the apparent Michaelis constant that would be determined experimentally using the known bulk substrate concentrations. If the substrate is negatively charged, the following relationships hold:

$$\Lambda = [S]/[S_o] \qquad (3.6)$$

$$K_m^{app} = K_m/\Lambda \qquad (3.7)$$

The K_m of an enzyme for a substrate is apparently reduced if the substrate concentration in the vicinity of the enzyme's active site is higher than that measured in the bulk of the solution (Figure 3.7). This is because a lower bulk concentration of the substrate is necessary in order to provide the higher localised substrate concentration needed to half-saturate the enzyme with substrate. Similar effects on the local concentration of products, inhibitors, cofactors and activators may change the apparent kinetic constants involving these molecules. For example, the apparent inhibition constant of a positively charged competitive inhibitor is given by

$$K_i^{app} = K_i\Lambda \qquad (3.8)$$

If the immobilised enzyme contains a number of groups capable of chelating cations, the partition of such cations into the microenvironment is far greater than that described by the electrostatic partition coefficient (Λ). For example, soluble glucose isomerase needs a higher concentration of magnesium ions than that required by the immobilised enzyme. It also requires the presence of cobalt ions, which do not need to be added to process involving immobilised enzyme due to their strong chelation by the immobilised matrix. A high concentration of ionising groups may cause partitioning of gases away from the microenvironment, with consequent effects on their

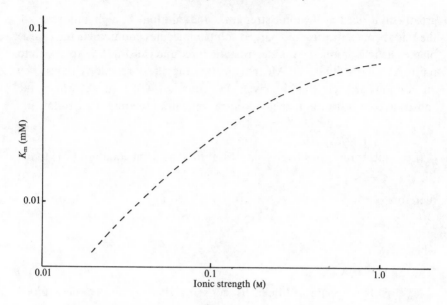

Figure 3.7. The effect of immobilisation and ionic strength on the K_m of bromelain for its positively charged substrate, N-α-benzoyl-L-arginine ethyl ester. The support is the negatively charged polyanionic polymer carboxymethyl cellulose (Engasser & Horvath, 1976). ——— Free enzyme; ------- immobilised enzyme.

apparent kinetic parameters. It is also a useful method for protecting oxygen-labile enzymes by 'salting out' the oxygen from the vicinity of the enzyme.

Differential partitioning of the components of redox couples may have a significant effect on the activity and stability of certain enzymes. For example, papain is stabilised by the presence of thiols, which act as effective reducing agents. However, thiols possess partial negative charges at neutral pH values, which causes their expulsion from the microenvironment of papain if it is immobilised on the negatively charged clay, kaolinite. In effect, the redox couple involving thiol and uncharged disulphide, which is not partitioned, becomes more oxidising around this immobilised enzyme. The net effect is a destabilisation of the immobilised papain relative to the free enzyme:

$$2H^+ + 2R\text{--}S^- \xrightleftharpoons{\quad pK_a \approx 8\quad} 2R\text{--}SH \rightleftharpoons R\text{--}S\text{--}S\text{--}R + 2H^+ + 2e^- \quad [3.6]$$
$$\underset{\text{(stabilised enzyme)}}{\underset{\text{reduced}}{}} \qquad \underset{\text{(destabilised enzyme)}}{\underset{\text{oxidised}}{}}$$

Partition of hydrogen ions represents an important case of soluble partition. The pH of the microenvironment may differ considerably from the pH of the bulk solution if hydrogen ions are partitioned into or out of the

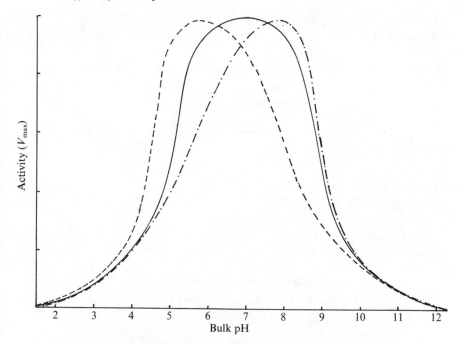

Figure 3.8. Diagram of the variation in the profiles of activity of an enzyme, immobilised on charged supports, with the pH of the solution. ——— Free enzyme. —————— Enzyme bound to a positively charged cationic support; a bulk pH of 5 is needed to produce a pH of 7 within the microenvironment. —·—·—·— Enzyme bound to a negatively charged anionic support; a pH of 7 within the microenvironment is produced by a bulk pH of 9.

immobilised enzyme matrix. The binding of substrate and the activity of the immobilised enzyme both depend on the local microenvironmental pH, whereas the pH as measured by a pH meter always reflects the pH of the bathing solution. This causes apparent shifts in the behaviour of the kinetic constants with respect to the solution pH (Figure 3.8). It is quite a simple process to alter the optimum pH of an immobilised enzyme by 1–2 pH units, giving important technological benefits (e.g. allowing operation of a process away from the optimum pH of the soluble enzyme but at a pH more suited to the solubility or stability of reactants or products).

In addition to its effects on solute partition, the localised electrostatic gradient may affect both the K_m and V_{max} by encouraging or discouraging the intramolecular approach of charged groups within the enzyme, or enzyme–substrate complex, during binding and catalysis. A large number of small energetic gains and losses may complicate the analysis of such overall effects (Table 3.4). As the resultant changes are also reduced by increases in the ionic

Table 3.4   *The effect of covalent attachment to a charged matrix on the kinetic constants of chymotrypsin for* N-acetyl-L-tyrosine ethyl ester *(Goldstein, 1972)*

	Ionic strength (M)	k_{cat} (s^{-1})	K_m (mM)	Specificity (k_{cat}/K_m)
Free enzyme	0.05	184	0.74	249
	1.00	230	0.55	418
Enzyme attached to a negatively charged support	0.05	300	2.50	120
	1.00	280	1.93	145
Enzyme attached to a positively charged support	0.05	119	7.10	17
	1.00	165	5.82	28

The changes in k_{cat} may be due to the approach of two positively charged groups during the rate-controlling step in the catalysis.

strength of the solution, these electrostatic effects may be difficult to distinguish from the effects of partition.

Hydrophobic interactions play a central role in the structure of lipid membranes and the conformation of macromolecules, including enzymes. They are responsible for the relative solubility of organic molecules in aqueous and organic solvents. These interactions involve an ordered rearrangement of water molecules at the approach of hydrophobic surfaces. The force of attraction between hydrophobic surfaces decays exponentially with their distance apart, halving with every nanometer of separation. These hydrophobic interactions effectively reduce the dielectric constant of the microenvironment with consequent modification of the acidity constants of acid and basic groups on the enzymes, substrates, products and buffers (Figure 3.9). Similar effects may alter the acidity constants of key substrate binding groups, so affecting the K_m of the immobilised enzyme for its substrates. Hydrophobic interactions are unaffected by the ionic strength or pH of the solution but may be neutralised by the presence of neighbouring hydrophilic groups, where they are sufficient to dominate the localised structure of the water molecules. Hydrophobic interactions may, therefore, cause the partition of molecules between the bulk phase and the microenvironment. If the surface of the immobilised enzyme particles is predominantly hydrophobic, hydrophobic molecules will partition into the microenvironment of the enzyme and hydrophilic molecules will be partitioned out into the

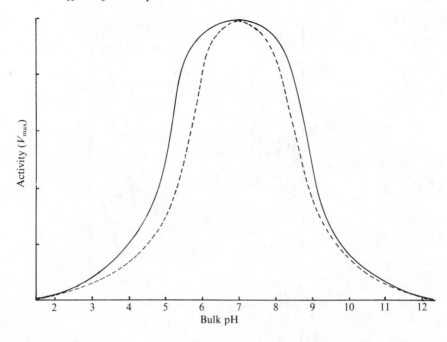

Figure 3.9. Diagram showing the effect of a hydrophobic support on the pH–activity profile of immobilised enzymes in solutions of low ionic strength. ———— Free enzyme; ——————— enzyme immobilised on a hydrophobic support. The effective decrease in the dielectric constant for the microenvironment reduces the dissociation of charged groups, increasing the pK_a of carboxylic acids and changing the pK_a of some other groups.

bathing solution. The reverse case holds if the biocatalytic surface is hydrophilic. Partition causes changes in the local concentration of the molecules which, in turn, affects the apparent kinetic constants of the enzyme in a manner similar to that described for immobilised enzyme particles possessing a net charge. An example of this effect is the reduction in the K_m of immobilised alcohol dehydrogenase for butanol. If the support is polyacrylamide, the K_m is 0.1 mM but, if the more hydrophobic co-polymer of methacrylate with acrylamide is used as the support, the K_m is reduced to 0.025 mM. In this example, no difference is noticed in the apparent values for K_m using ethanol, a more hydrophilic substrate. A similar effect may be seen in the case of competitive inhibitors (Table 3.5). Gases (e.g. oxygen) partitioned out from the microenvironment by the presence of a charged support are generally partitioned into the microenvironment by hydrophobic supports.

Other specific partition effects are associated with particular immobilisation supports. For example: (*a*) the apparent K_m of glucoamylase for

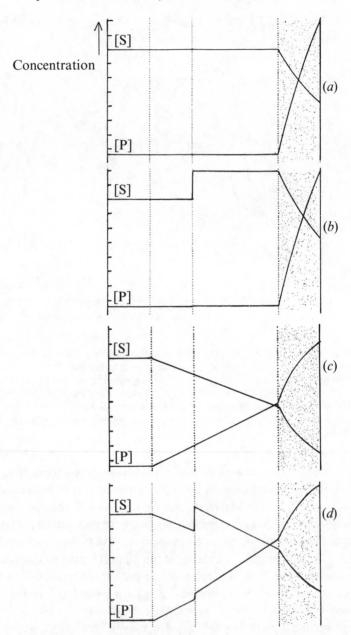

Figure 3.10. Diagram showing the concentration gradients of substrate and product that may be produced around a porous particle of an immobilised enzyme. (*a*) Concentration gradient due solely to reaction and internal diffusion within the particle; (*b*) as (*a*) but with additional concentration gradients due to partition of substrate and product into the microenvironment; (*c*) as (*a*) but with an additional concentration

Table 3.5   *The effect of immobilisation using an hydrophobic support on the relative competitive inhibition of invertase*

	Invertase	
Inhibition constant (K_i, mM)	Soluble	Bound to polystyrene (hydrophobic)
Aniline (hydrophobic)	0.94	0.39
Tris-(hydroxymethyl) aminomethane (hydrophilic)	0.45	1.10

The K_i is reduced where both the inhibitor and support are hydrophobic.

maltose is considerably reduced when the enzyme is immobilised on titanium-activated supports, such supports having a specific affinity for some polyalcohols; (b) some polyphenolic resins have specific affinities for polysaccharides which assist their partition into the microenvironment.

Effects of solute diffusion on the kinetics of immobilised enzymes

In order for an immobilised enzyme to catalyse a reaction, the substrates must be able to diffuse through the solution to the catalytically active sites and the products must be able to diffuse away into the bulk solution. The driving force for the net diffusive process is due to the concentration gradients, solutes moving in the direction of higher to lower concentration. The substrates approach the surface of the enzyme particles through the surrounding thin stagnant unstirred layer of solution and then diffuse into any pores where they may encounter active enzyme. The net movement of the solutes is described in terms of the following two steps: (1) *external diffusion*, where the transport of substrates towards the surface, and products away from it, is in series with the catalytic conversion of substrates to products

Figure 3.10 (*cont.*)
gradients due to external diffusion to the surface of the particle; (d) concentration gradients due to the combined effects of partition and diffusion. The partition boundary layer is normally about a thousand-fold thinner than the diffusive boundary layer. The actual concentration gradients will not show the sharp discontinuities which are shown here for simplicity.

occurring at the surface (the processes being consecutive); and (2) *internal diffusion*, where the transport of the substrates and products, within the pores of immobilised-enzyme particles, is in parallel with the catalysed reaction (the processes being simultaneous). The concentration gradients caused by diffusion and partition are shown diagrammatically in Figure 3.10. The rate of a reaction catalysed by an immobilised enzyme (v) is normally lower than the rate due to the same amount of free enzyme in solution (v_{free}). This is due to the controlling necessity for the substrate to diffuse from the bulk phase to the catalytic surface. The substrate concentration within the microenvironment ($[S]$) is lower than that in the bulk ($[S_o]$) due to its depletion by the reaction. The change in reaction rate can be expressed quantitatively by introducing the *effectiveness factor* (η), where:

$$\eta = v/v_{free} \tag{3.9}$$

The effectiveness factor generally lies between 0 and 1 and is dependent on the bulk substrate concentration, amongst other factors (see later). It may sometimes be greater than unity, due to non-isothermal operation, because of partition or inhibitory effects, or if the immobilised enzyme is stabilised relative to the free enzyme over the time-course of its assay.

The effect of external diffusion on the rate of an enzyme-catalysed reaction may be simply derived, assuming (1) that Michaelis–Menten kinetics operate, (2) that the enzyme is immobilised to a flat impervious support, and (3) that there is an absence of partitioning or electrostatic effects. If the reaction is occurring under steady-state conditions, the rate of increase of product within the bulk of the solution must equal the rates of three consecutive processes; the rate at which substrate diffuses to the surface, the rate of enzymic catalysis, and the rate at which the product diffuses away from the surface. The steady-state assumption is generally valid if the volume of the bulk of the solution is sufficiently large such that the variation in $[S_o]$ with time may be ignored. Immobilised enzymes are open systems, where both energy and material are exchanged through the boundary with the environment. This allows steady-state operation even at very high enzyme loading. In any circumstances it is clear that the substrate concentration at the catalytic surface cannot continuously increase or decrease at a substantial rate compared with the rate of reaction.

Therefore, from equation (1.8) (p. 9):

$$v = \frac{A V_{max}[S]}{K_m + [S]} \tag{3.10}$$

where A is the total surface area and V_{max} is the maximum rate of reaction catalysed by unit area of surface. Combining equations (1.8) and (3.9) gives:

$$v = \frac{\eta A V_{max}[S_o]}{K_m + [S_o]} \tag{3.11}$$

As η varies with $[S_o]$, the immobilised enzyme will no longer show hyperbolic kinetics. Eadie–Hofstee plots (see Figure 3.18) derived from the relationship (3.11) are not linear because they obey the transformation equation:

$$v/V_{max} = \eta A - (v/V_{max})(K_m/[S_o]) \tag{3.12}$$

As this also involves a term in η which varies, equalling unity where v equals V_{max} at the intercept with the ordinate (v axis) but which may be much lower when v is much less than V_{max} at the intercept with the abscissa ($v/[S_o]$ axis).

Assuming that all of the surface is equally accessible, the rate of flow of substrate to the surface has been found to be proportional both to the surface area and to the difference in substrate concentration between the bulk of the solution and the microenvironment next to the surface. It is given by the relationship:

$$v = k_L A([S_o] - [S]) \tag{3.13}$$

The proportionality constant (k_L) is known as the (local liquid phase) *mass transfer coefficient* (with units of m s^{-1}), which depends upon the diffusivity of the substrate and the effective distance between the surface and the bulk phase.

$$k_L = D_S/\delta \tag{3.14}$$

where D_S is the substrate diffusivity (diffusion coefficient) in free solution (with units of m^2 s^{-1}) and δ is the effective thickness of the unstirred layer through which the substrate must diffuse. The diffusivity is defined by Fick's law as the rate at which a unit mass (m) of the compound travels through a unit surface area (A) due to a concentration gradient of unit change in density (ρ):

$$\frac{dm}{dt} = -D_S A \frac{d\rho}{dx} \tag{3.15}$$

D_S depends upon the molecular weight and dimensions of the substrate, and the temperature, viscosity and composition of the liquid phase (Table 3.6). In general terms, the higher the molecular weight and the solution viscosity, the lower will be the diffusivity. δ may be effectively regarded as a distance, although it is not precisely defined by the thickness of the unstirred stagnant layer, as liquid motion may be detectable at distances less than δ from the surface. Under stagnant flow conditions, δ is equal to the particle radius for spherical particles. It depends on the hydrodynamic conditions, being reduced by increases in the rate of stirring and consequent increasing particle-fluid relative velocity, this reduction in δ causing an increase in the

Table 3.6   *Diffusivity of molecules in aqueous solution at 20 °C*

Substance	Molecular weight	Diffusivity $(m^2 s^{-1} \times 10^{10})$
Oxygen	32	21.0
Glucose	180	6.7
Sucrose	342	4.5
Inulin	5200	2.3
Albumin	67000	0.7
Urease	480000	0.3
Bushy stunt virus	10700000	0.1

mass transfer coefficient (k_L). For example, using particles of 400 μm diameter, δ has been found to be 5μm when used in a packed bed reactor (see Chapter 5) at a reactant stream flow rate of 1 m s^{-1}. Depending upon the conditions, δ also may vary with the diffusivity, and the density and viscosity of the liquid, increasing with increasing diffusivity and viscosity but reducing with increasing density difference between the immobilised biocatalyst and the medium. It may be halved by the use of ultrasound, which may be particularly useful for larger particles. For the small biocatalytic particles normally used, the maximum reduction in δ that is achievable by increasing the turbulence of the solution around the immobilised enzyme is about 10-fold.

The rate of diffusion is significantly affected where partitioning occurs. For charged molecules, this will depend on the electrostatic potential gradient in addition to the concentration gradient. The electrostatic potential gradient causes apparent changes in both δ and D_S within the immediate vicinity of the surface (i.e. at distances very much smaller than δ). This may be particularly relevant in the case of hydrogen ion diffusion (see later) as hydrogen ions move rapidly through solutions across electrostatic gradients, due to their ability rapidly to change their association with water molecules.

Due to the consecutive nature of the process the rate of enzymic reaction (given by v in equation (3.10)) must equal the rate of diffusion of the substrate to the catalytic surface (given by v in equation (3.13)), the terms in A cancelling out:

$$\frac{V_{max}[S]}{K_m + [S]} = k_L([S_o] - [S]) \tag{3.16}$$

This equation is quadratic with respect to the microenvironmental substrate concentration ([S]), the value of which is very difficult to establish by independent means. It may be simplified under extreme values of [S] relative to the K_m of the enzyme for the substrate.

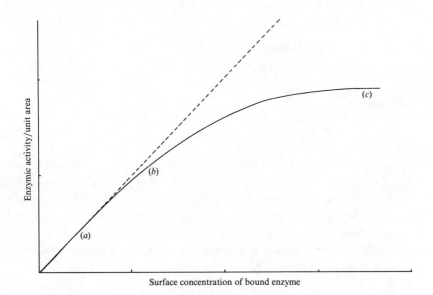

Figure 3.11. Variation, with its concentration, of the rate of reaction catalysed by an immobilised enzyme. The relationship shows three phases: (*a*) kinetic control by the enzyme, extrapolated (------) to show the activity of equivalent amounts of free enzyme; (*b*) mixed intermediate control; (*c*) control by the rate of external transport of substrate.

If [S] is much greater than K_m, the left-hand side of equation (3.16) simplifies to give just V_{max}:
therefore:

$$V_{max} = k_L([S_o] - [S]) = v_A \qquad (3.17)$$

where v_A is the rate of reaction catalysed by unit area of the immobilised enzyme surface. It follows that the rate of the reaction is equal to the maximum rate of reaction of the non-immobilised enzyme when [S], and hence [S_o], is much greater than the K_m. Often, however, it is found that [S] is much less than K_m. Under such conditions, equation (3.16) gives:

$$[S](V_{max}/K_m) = k_L([S_o] - [S]) \qquad (3.18)$$

Collecting terms in [S]:

$$[S] = \frac{k_L[S_o]}{k_L + V_{max}/K_m} \qquad (3.19)$$

therefore:

$$[S] = \frac{[S_o]}{1 + (V_{max}/K_m)(1/k_L)} \qquad (3.20)$$

From equation (3.10), when [S] is much less than K_m:

$$v_A = [S](V_{max}/K_m) \tag{3.21}$$

Substituting for [S] from equation (3.20):

$$v_A = \frac{[S_o](V_{max}/K_m)}{1 + (V_{max}/K_m)(1/k_L)} \tag{3.22}$$

therefore:

$$v_A = \frac{[S_o]}{1/k_L + 1/(V_{max}/K_m)} \tag{3.23}$$

The relative values of the two components of the denominator in equation (3.23) determine whether the reaction is controlled primarily by the diffusion of the substrate (the k_L term) or by the catalytic ability of the immobilised enzyme (the V_{max}/K_m term). The comparison may be made by means of the introduction of an external substrate modulus (μ, also known as the *Damköhler number* and is the dimensionless ratio of reaction velocity to transport velocity) defined by:

$$\mu = \frac{V_{max}}{k_L K_m} \tag{3.24}$$

where V_{max} is the maximum rate of reaction catalysed by unit surface area. Substituting this in equation (3.22):

$$v_A = \frac{[S_o](V_{max}/K_m)}{1 + \mu} \tag{3.25}$$

The relationships between the rates of reaction and the immobilised enzyme and bulk substrate concentrations are shown in Figures 3.11 and 3.12. When k_L is much greater than V_{max}/K_m (i.e. at zero μ, when mass transport is capable of a much faster rate than that of the enzyme-catalysed reaction), the overall rate of the process is under the kinetic control of the enzyme, which is then as effective as the free enzyme (i.e. $\eta = 1$). This allows the simplification of equation (3.23) to yield:

$$v_A = [S_o](V_{max}/K_m) \tag{3.26}$$

However, when k_L is much less than V_{max}/K_m (e.g. at high substrate modulus, when mass transport is much slower than the intrinsic rate of the enzyme-catalysed reaction), the overall rate of the process is under the control of the rate of diffusion of the substrate. Equation (3.23) then simplifies to give:

$$v_A = k_L[S_o] \tag{3.27}$$

This last relationship, apart from its obvious utility as a method for determining mass transport coefficients, is very important for the proper understanding of the behaviour of immobilised enzymes under the diffusio-

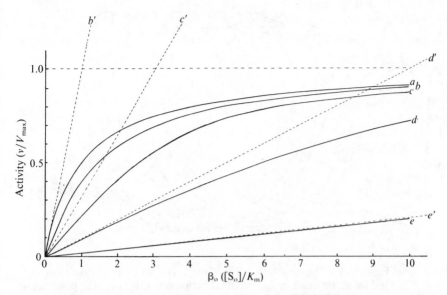

Figure 3.12. The variation in the rate of reaction catalysed by an immobilised enzyme and the dimensionless bulk substrate concentration (β_o, which equals $[S_o]/K_m$) with the external substrate modulus (μ, defined by equation (3.24)). (Curve a) Free enzyme; (curve b) $\mu = 1$; (curve c) $\mu = 3$; (curve d) $\mu = 10$; (curve e) $\mu = 100$. Also drawn (------) are the maximum rates of substrate diffusion to the surface, given by $v_A/V_{max} = k_L\beta_o$ (lines b', c', d' and e' showing the effect of decreasing mass transfer coefficients, $k_L = 1.00, 0.33, 0.10$ and 0.01, corresponding to curves b, c, d and e, respectively). It should be noticed that curves b, c, d and e, are bounded both by curve a and by the lines b', c', d' and e', respectively. Increased diffusional control extends the range of linearity at lower substrate concentrations, but the same V_{max} is reached in all cases if a sufficiently high substrate concentration can be achieved (see Figure 3.27 (a)).

nal control of external transport of the substrate. The rate of reaction is shown to be independent of the activity of the enzyme. This means that it is not affected by changes in the pH, temperature (except as it may affect viscosity) or ionic strength of the solution, nor is it affected by the presence of inhibitors or activators. If, however, the ratio V_{max}/K_m is reduced substantially by changes in these conditions to approach the value of k_L, the relationship shown in equation (3.27) will no longer hold true. As diffusional limitations approach a total controlling influence, the behaviour of the enzyme with respect to these factors gradually changes from that of a free enzyme in solution to the state of being unaffected. For example, increasing diffusional control causes a broadening of the pH–activity profile and a lowering of the activation energy (and the related Q_{10}), both of which will be

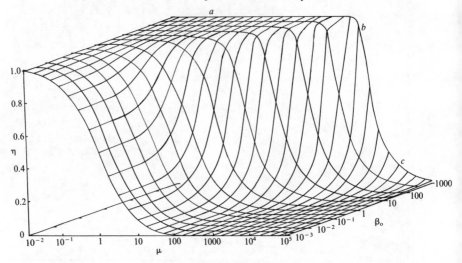

Figure 3.13. The combined effect of the bulk substrate concentration (β_o) and substrate modulus (μ) on the effectiveness factor (η). The plateau (*a*) is an area of kinetic control, the surface dropping through an area of intermediate control (*b*) to an area of diffusional control (*c*).

more apparent at lower substrate concentrations. Under such conditions, the bulk substrate concentration that gives half the maximum rate of reaction ($[S_{1/2}]$, equivalent to K_{app}^m) is higher than the K_m of the free enzyme (see Figure 3.12). This is in contrast to the case of the free enzyme, where $[S_{1/2}]$ is identical with the K_m. The equivalent microenvironmental substrate concentration giving half-maximal reaction rate remains equal to the K_m.

v_A may be substituted by $V_{max}/2$ in equation (3.13) when $[S]$ equals K_m, giving:

$$v_A = V_{max}/2 = k_L([S_{1/2}] - K_m) \tag{3.28}$$

Therefore:

$$[S_{1/2}] = K_m + V_{max}/2k_L \tag{3.29}$$

Therefore:

$$[S_{1/2}] = K_{app}^m = K_m(1 + \mu/2) \tag{3.30}$$

The K_m apparently increases with the external substrate modulus. This causes a reduction in the apparent specificity. The introduction of the dimensionless substrate concentrations $\beta\,(= [S]/K_m)$ and $\beta_o\,(= [S_o]/K_m)$ into equation (3.16) gives:

$$\frac{V_{max}\beta}{1 + \beta} = k_L K_m(\beta_o - \beta) \tag{3.31}$$

Substituting μ from equation (3.24):

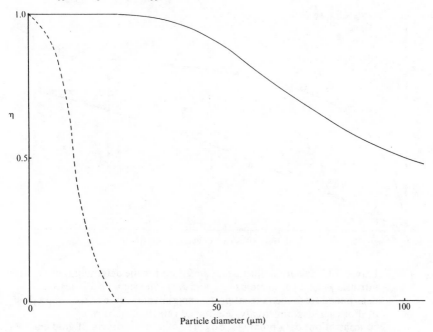

Figure 3.14. Diagram showing the effect of particle diameter on the effectiveness factor (η) of immobilised enzymes. ——— Small surface concentration of enzyme; —————— high surface concentration of enzyme.

$$\frac{\mu\beta}{1 + \beta} = \beta_o - \beta \qquad (3.32)$$

This may be simplified at low β_o, when β approaches zero:

$$\beta = \beta_o/(1 + \mu) \qquad (3.33)$$

and:

$$[S] = [S_o]/(1 + \mu) \qquad (3.34)$$

Therefore, if the value of μ is known, the concentration of the substrate at the surface of the immobilised enzyme may be obtained.

The most important factors to arise from this analysis concern the consequences of immobilisation on the effective catalytic ability of the enzyme. It is clear that the effectiveness factor (η) must vary with both β_o and μ, being reduced by low β_o and high μ; this relationship is illustrated in Figure 3.13. The conditions that produce an increased probability of external diffusional control over the rate of an immobilised-enzyme catalysed reaction may be summarised as follows:

(1) High enzyme loading on the surface.
(2) Low bulk substrate concentration.

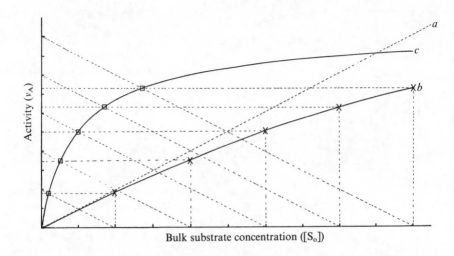

Figure 3.15. Diagram illustrating a method for the determination of the intrinsic kinetic parameters (V_{max} and K_m). The value of k_L may be determined from the tangent of the experimental curve at the origin (line *a*). For each experimentally determined rate:bulk substrate concentration) data point (×, on curve *b*) a line is drawn of gradient $-k_L$ from the appropriate position on the horizontal axis. The microenvironmental substrate concentration is given by the intercept of this line with the vertical position representing the reaction rate. These graphically determined (rate:microenvironmental substrate concentration) data points (□, on curve *c*) may be used to calculate the intrinsic kinetic parameters.

(3) Low substrate diffusivity.

(4) Low K_m.

(5) High enzymic specificity.

(6) Low rate of stirring or mixing.

(7) Flat surfaces (e.g. large average particle diameter; see Figure 3.14).

Diffusion-free enzyme kinetics can be simply determined by decreasing the loading of enzyme on the immobilisation support or by lowering the temperature. It is often necessary to determine the kinetic parameters of an immobilised enzyme in the presence of external diffusional effects. This may be in order to investigate the effect of immobilisation on the intrinsic stability or activity of the enzyme. The intrinsic V_{max} can be determined if sufficiently high microenvironmental substrate concentrations can be achieved (see Figure 3.27(*a*)) but determination of the intrinsic K_m depends upon knowledge of the microenvironmental substrate concentration. This may be determined graphically if the mass transfer coefficient is known (Figure 3.15).

The substrate concentration gradient within the microenvironment is only

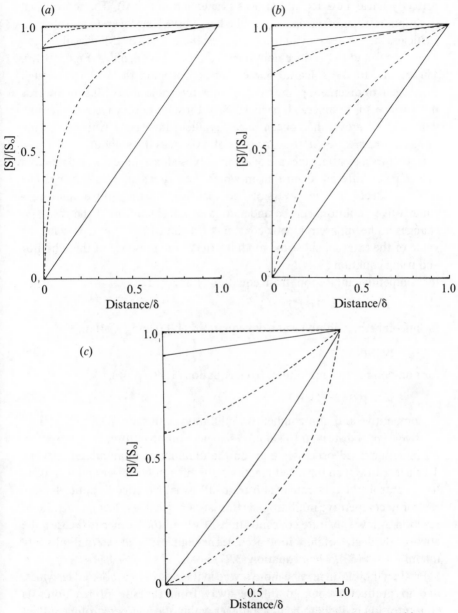

Figure 3.16. Concentration profiles of substrate molecules approaching the curved surface of immobilised enzymes, where r is the radius of curvature. (*a*) Enzyme attached to spherical particles; (*b*) enzyme attached to the outside of cylindrical fibres; (*c*) enzyme attached to the inside of cylindrical tubes. ———— $r/\delta = 9$; –––––– $r/\delta = 0.11$. In each case, $\mu = 10^{-3}$, 1 and 10^3 for the top, middle and bottom pairs of curves, respectively.

perfectly linear next to flat surfaces (see equation (3.13)). The more usual situation is that of enzymes attached to curved surfaces (e.g. spherical particles, the inside of cylindrical tubes or the outside of cylindrical fibres). These are found to produce non-linear concentration profiles (Figure 3.16). This is due to the substrate molecules approaching the surface through convergent or divergent pathways (e.g. substrate molecules diffusing towards the surface of a spherical particle pass through successively decreasing volumes and areas, the concentration gradient increasing with decreasing radius in order to retain the same flux of molecules throughout).

Inhibitors may affect the rate of an immobilised-enzyme catalysed reaction in a manner different from that in which they affect the free enzyme. The relative effects of the reversible competitive, uncompetitive and non-competitive inhibitors can be understood by consideration of the resultant changes in the apparent kinetic constants K_{app}^m and V_{app}^{max}. If μ_i represents the value of the external substrate modulus (μ) in the presence of the inhibitor and using equation (3.24):

for competitive inhibition, from equation (1.84) (p. 29):

$$\mu_i = \mu/(1 + [I]/K_i) \tag{3.35}$$

for uncompetitive inhibition, from equations (1.93) and (1.94) (p. 32):

$$\mu_i = \mu \tag{3.36}$$

for non-competitive inhibition, from equation (1.98) (p. 36):

$$\mu_i = \mu/(1 + [I]/K_i) \tag{3.37}$$

Competitive and non-competitive inhibition both show a reduction in μ_i relative to μ. Substrate diffusional resistance will, therefore, have less effect on these inhibited processes than on the uninhibited immobilised enzyme. This is because such inhibited reactions are inherently slower and less likely to be controlled by the rate of substrate diffusion. No effect is noticed in the case of uncompetitive inhibition, as this has a negligible effect on the rate of reaction at low substrate concentrations. Even in the former two cases, the effect of the degree of inhibition becomes negligible at high external substrate modulus ($\mu > 50$), when equation (3.27) holds.

Product inhibition may be more severe in the case of immobilised enzymes, due to product having to diffuse away from the site of reaction. Its concentration is likely to be much higher within the microenvironment than in the bulk of the solution (Figure 3.10). At the beginning of the reaction, this product concentration will build up until the concentration gradient to the bulk macroenvironment is sufficient to allow it to diffuse away at a rate equal to that of its production by the reaction. If k_L^S and k_L^P represent the substrate and product mass transfer coefficients, and [P] and [P_o] represent the product

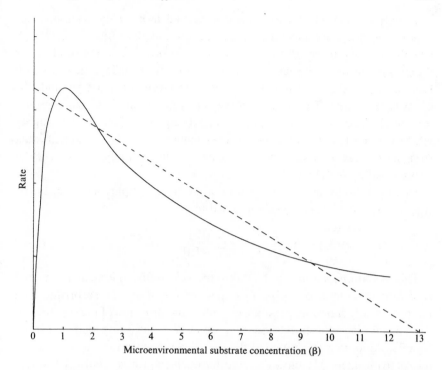

Figure 3.17. The effect of substrate inhibition on the rate of immobilised enzyme reactions. ——— Immobilised enzyme versus microenvironmental substrate concentration (or free enzyme versus bulk substrate concentration); -------- the rate of diffusion of the substrate from the bulk ($\beta_o = 13$) to the microenvironment (i.e. rate $= k_L(\beta_o - \beta)$, lower microenvironmental substrate concentration, relative to the bulk concentration, giving the higher rate.

concentrations in the microenvironment and bulk, respectively, then from equation (3.13):

$$v_A = k_L^S([S_o] - [S]) = k_L^P([P] - [P_o]) \tag{3.38}$$

Under diffusional control, [P] and $[S_o]$ are much greater than $[P_o]$ and [S], respectively:

Therefore:

$$[P] = [S_o]k_L^S/k_L^P \tag{3.39}$$

Substituting for [P] in equation (1.85) (p. 29) and assuming that K_m is much greater than [S]:

$$v_A = \frac{V_{max}[S]}{K_m + K_m k_L^S[S_o]/(K_P k_L^P)} \tag{3.40}$$

The effect of product inhibition (the second term in the denominator) depends on the bulk substrate concentration and the ratio $K_m k_L^S/(K_P k_L^P)$, which expresses the competition between the substrate and the product for the enzymic surface. The build-up of the product at the surface increases with this ratio, causing a reduction in both the rate of reaction and the effectiveness factor. The effect is greater when K_m/K_p or k_L^S/k_L^P are large and μ is small (< 50). k_L^S is usually approximately equal to k_L^P but may be higher where the reaction involved produces a higher molecular weight product from more than one substrate molecule, or substantially lower in a depolymerisation (e.g. hydrolysis) reaction.

Substrate inhibition presents a somewhat more complex scenario. Equations (1.96) and (3.13) may be combined to give:

$$v_A = k_L([S_o] - [S]) = \frac{V_{max}[S]}{K_m + [S] + [S]^2/K_S} \qquad (3.41)$$

This equation is third order with respect to the microenvironmental substrate concentration ([S]). For this reason, it is not surprising that multiple steady states are possible, provided that diffusion of the substrate to the enzyme is sufficiently slow (Figure 3.17). The same bulk substrate concentration may give two different stable concentrations within the microenvironment: (1) a low concentration which gives a relatively fast rate of reaction, without much substrate inhibition, and equally fast rate of inward diffusion of substrate due to the steep concentration gradient; (2) a much higher concentration which gives a relatively slow rate of reaction, due to the substrate inhibition, and equally slow rate of inward diffusion of substrate down the relatively slight concentration gradient. A third possibility exists of an intermediate concentration, representing an unstable state which is not naturally established and is of no practical consequence (Figure 3.17). The choice of which stable state exists depends on the start-up conditions. Addition of the substrate under conditions of low external substrate modulus (e.g. at a low temperature or with vigorous stirring) allows an initially high microenvironmental substrate concentration to be achieved. A stable state is favoured that involves higher [S] as the temperature and external substrate modulus are raised. The alternative stable state may be reached by addition of the substrate under conditions of high external substrate modulus, where the microenvironmental substrate concentration is initially zero and kept low.

Reversible reactions catalysed by immobilised enzymes may be severely affected by the slow diffusion of the product away from the catalytic surface. Even a slight build-up in the microenvironmental product concentration will increase the reverse rate of reaction, severely reducing the productivity of the enzyme. An analysis of the effect of this diffusional resistance may be made by combining equations (1.68) (p. 26) and (3.38), which then describe the resultant lowering of the rates of reaction and effectiveness factor. Figure 3.18 shows Eadie–Hofstee plots comparing the effect of reaction reversibility at increasing external substrate modulus.

Diffusional resistance can lead to an apparent increase in the stability of immobilised enzymes. The reason for this is the ceiling that the diffusional resistance imposes upon the activity. At a high loading of active enzyme, the activity obeys equation (3.27) and is independent of the inherent activity of the immobilised enzyme. Under these circumstances, there is always sufficient reaction occurring to remove the substrate as it arrives from the bulk solution, most of the enzyme molecules being effectively redundant (effectiveness factor close to zero). The productivity of the immobilised biocatalyst remains constant until the specific activity of the enzyme has decayed sufficiently that the reaction is no longer diffusionally controlled (Figure 3.19). Some of the early researchers in this field who used data collected only during the initial diffusionally controlled period were misled into believing that the immobilisation process was more likely than is now realised to stabilise an enzyme.

A special case of diffusional control of immobilised-enzyme reactions concerns hydrogen ions. Many enzyme-catalysed reactions involve the release or consumption of H^+ (e.g. dehydrogenases, peptidases and esterases). This may include any reaction involving molecules containing ionisable groups as the pK_a value of such groups on the substrates and products may differ. Sometimes the reaction may produce or consume hydrogen ions, depending on the pH of the reaction (e.g. urease, see Table 6.2, p. 207). Hydrogen ions diffuse relatively slowly, similar to other monovalent ions, in the absence of an electrical field. This is due to four main causes: (1) H^+ is normally hydrated (H_3O^+, $H_9O_4^+$) increasing its apparent molecular weight and hence reducing its diffusion coefficient; (2) the need for localised electroneutrality necessitates the co-diffusion of positively and negatively charged species, again causing an increase in the effective molecular weight (i.e. each H^+ diffuses with an anionic counter-ion); (3) hydrogen ions are buffered by histidine and other groups on the immobilised-enzyme particles which reduce their ability to diffuse into the bulk liquid phase; (4) normally the H^+ concentration in the bulk of the solution is low (e.g. pH 7 is equivalent to 10^{-7} M H^+), which allows the production of only very small concentration gradients, even at pH differences of 1 or 2. The shallowness of H^+ concentration gradients are especially noticeable when compared with those normally encountered for other substrates or products (greater than mM). For these reasons many reactions may be controlled by the diffusion of

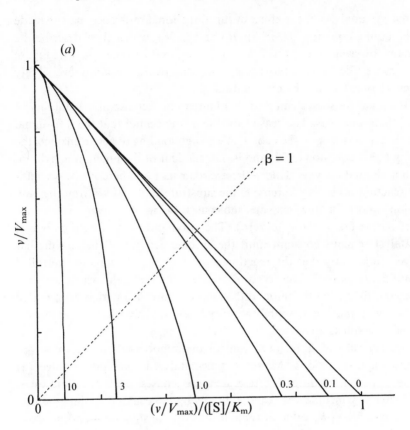

(a)

v/V_{max}

$\beta = 1$

10   3   1.0   0.3   0.1   0

$(v/V_{max})/([S]/K_m)$ 1

hydrogen ions to, or away from, immobilised-enzyme surfaces. The effect of this diffusional control may be examined by combining equations (1.17) (p. 15) and (3.13) for a reaction involving the production of H^+:

$$\frac{V^*}{([H^+]/K_{a1}) + 1 + (K_{a2}/[H^+])} = k_L^H([H^+] - [H_o^+]) \qquad (3.42)$$

where V^* is the maximum rate of reaction if pK_{a1} and pK_{a2} are well separated, k_L^H is the mass transfer coefficient of H^+, $[H^+]$ is the microenvironmental concentration of H^+, and $[H_o^+]$ is the bulk concentration of H^+. If the reaction involves the consumption of hydrogen ions, the reaction obeys the following related equation:

$$\frac{V^*}{([H^+]/K_{a1}) + 1 + (K_{a2}/[H^+])} = k_L^H([H_o^+] - [H^+]) \qquad (3.43)$$

The magnitude of the diffusional resistance is given by the *proton modulus* (μ_H), where:

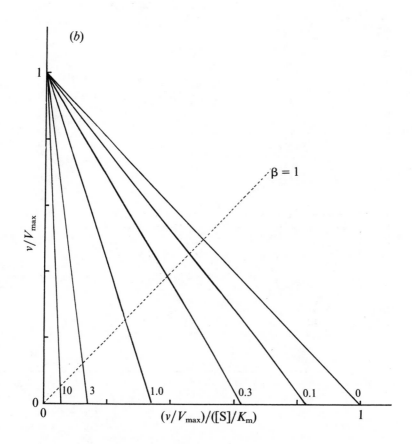

Figure 3.18. Eadie–Hofstee plots illustrating the effect of external diffusional resistance on the kinetics of an immobilised enzyme. Essentially non-reversible (*a*) and reversible (*b*) reactions are shown. The reversible situation has been modelled on the glucose isomerase reaction, using values of 1.14 for the K_{eq} and 42% and 51% (w/w, of carbohydrate present) for the initial bulk concentration of fructose and glucose, respectively. The lines represent the kinetically controlled reaction and increasing external substrate modulus (μ), as shown. Part (a) corresponds to Figure 3.12.

$$\mu_H = \frac{V^*}{k_L^H K_{a1}} \tag{3.44}$$

Note the similarity in the definitions of the proton and external substrate moduli (equation (3.24)).

It is found that hydrogen ions accumulate at the surface when the proton modulus is greater than about 10^{-4} (Figure 3.20). Such reactions may be

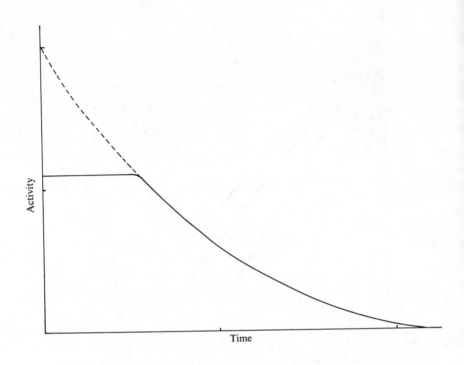

Figure 3.19. The apparent stabilisation of an immobilised enzyme, when diffusionally controlled. ———— Utilisable activity; ————— intrinsic activity, not realised above the activity ceiling imposed by the slower (rate-controlling) diffusion of the substrate to the enzyme.

operated at a significant rate, at a pH value well away from the optimum for the free enzyme. This may be very useful in cases where such a pH allows higher substrate solubilities or a more favourable process environment for the reaction. The diffusion of hydrogen ions may be facilitated by conjugate acid–base pairs, even at concentrations appreciably lower than those needed for their conventional buffering action. The reason for this is that the 'buffers' are capable of far greater concentration gradients than hydrogen ions, without necessarily affecting the bulk pH (Figure 3.21). In this case, equation (3.42) must be extended to include the diffusional transport of hydrogen ions by the conjugate acid. For a reaction producing H^+ this becomes:

$$\frac{V^*}{([H^+]/K_{a1}) + 1 + (K_{a2}/[H^+])} = k_L^H([H^+] - [H_o^+])$$

$$+ k_L^{HB}([HB] - [HB_o]) \quad (3.45)$$

where k_L^{HB} is the mass transfer coefficient of HB, [HB] is the microenvironmental concentration of the conjugate acid (HB) and $[HB_o]$ is the bulk concentration of HB. At a bulk pH of 7, about a thousand times more hydrogen ions are transported by a millimolar buffer with a pK_a of 7 than as free hydrogen ion. The presence of these buffering ions can significantly affect the pH–activity profile of reactions limited by hydrogen ion diffusion (Figure 3.22).

Analysis of diffusional effects in porous supports

So far, only external diffusional control of immobilised-enzyme catalysed reactions has been described. This has been more due to the ease of the analysis than to the scale of its application. Although it is undoubtedly extremely important, most immobilised-enzyme catalysts are porous to some extent, which necessitates the examination of the effect of internal diffusional resistance. Porous particles generally have very large surface areas of up to several hundred square metres per gram enabling very high immobilised-enzyme loading. Where such biocatalysts operate under total external diffusional control (i.e. equation (3.27) holds) then the surface substrate concentration is effectively zero, no substrate is available for penetration into any pores and internal diffusion may be safely ignored as being of no importance. However, in all other cases it may well be relevant. This is because the pathways through which the substrate travels internally within the particles are generally much greater than those involved in external diffusional control (i.e. the pore length is several orders of magnitude greater than the depth of the surrounding stagnant layer, δ). It should be noted that the effect of external diffusional limitations can be moderated, and sometimes removed completely, by changing the flow rate of the substrate solution over the biocatalytic surface, but such changes have no influence on the diffusion of the substrate within the protected environment of the interior of the particles.

Diffusion of the substrate and product inside a porous biocatalyst occurs in parallel with the catalysed reaction. The more the enzyme-catalysed reaction reduces the substrate concentration within the particles, the greater will be the substrate concentration gradient created between the internal microenvironment and the bulk of the solution. This, in turn, will increase the rate at which the substrate is delivered to the enzyme molecules towards the outside of the particles, increasing their effectiveness. The productivity of

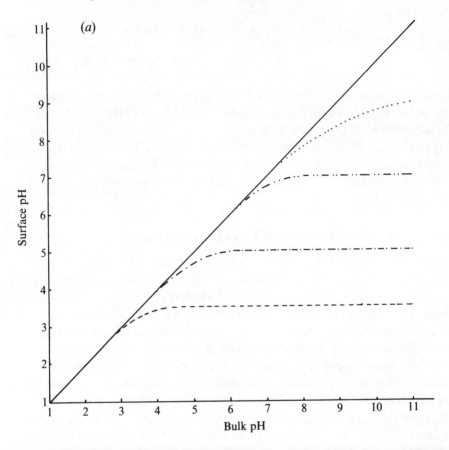

(a)

the reaction is, therefore, reduced considerably by the serious depletion of the substrate deep within the particle and the consequent high concentrations of product to be found there, possibly causing inhibition or reversing the reaction. To a certain extent, however, this is compensated for by the increased flow of substrate to the outer portion of the immobilised-enzyme particles due to the increased substrate concentration gradient.

Analysis of the effect of internal diffusion is complicated by such factors as: the shape of the particles, the distribution in size and shape of the pores, the total volume of the pores with respect to the particle volume (*porosity*, ϵ), the depth to which the pores penetrate the particles (e.g. pellicular particles have only a thin layer of enzyme-containing pores at their surface), the *tortuosity* (τ) of the route through the pores that the substrate encounters, the effective diffusivity of the reactants and products within the pores, and the degree of uniformity of the enzyme's distribution within the particles. (Many immobilisation methods produce a higher volumetric concentration of

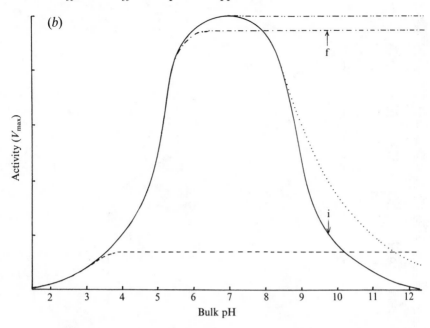

Figure 3.20. The effect of diffusional control on the local pH and pH–activity profile of an immobilised-enzyme catalysed reaction. (*a*) The variation of surface pH with bulk pH at various proton moduli (μ_H); ——— $\mu_H < = 10^{-5}$; ...... $\mu_H = 10^{-4}$; –··–··–·· $\mu_H = 10^{-3}$; –·–·–·– $\mu_H = 10^{-1}$; ––––– $\mu_H = 10$. The reduction in microenvironmental pH is most noticeable at high pH due to the much lower hydrogen ion concentration gradients, and hydrogen ion diffusion rates, present (e.g. a pH difference of 1 at pH 12 produces a concentration gradient only a millionth of that produced at pH 6). (*b*) The effect of the surface pH on the pH–activity profile. Note that μ_H and hence the pH–activity profile will vary with the flow rate and degree of turbulence. If the reaction is started at pH = 10 and $\mu_H = 0.1$, the rate of hydrogen ion production is initially low (initial rate at i) but faster than the rate that the hydrogen ions can diffuse away. This causes a drop in the microenvironmental pH towards the optimal pH, increasing the rate of reaction. This process continues until the concentration gradient between the microenvironmental and bulk phases is such that the rate of diffusion equals the rate of reaction (final rate at f).

enzyme towards the exterior of the immobilised enzyme particles, due to the rapid nature of the immobilisation reaction, which immobilises the enzyme before it penetrates the pores fully.) In order to examine the effect of these factors on a real system, it is useful to start with an analysis of the effect of internal diffusional resistance on the productivity of a 'model' porous biocatalytic particle. Beaded pellets are the most commonly encountered porous biocatalysts and may be considered as perfectly spherical for this

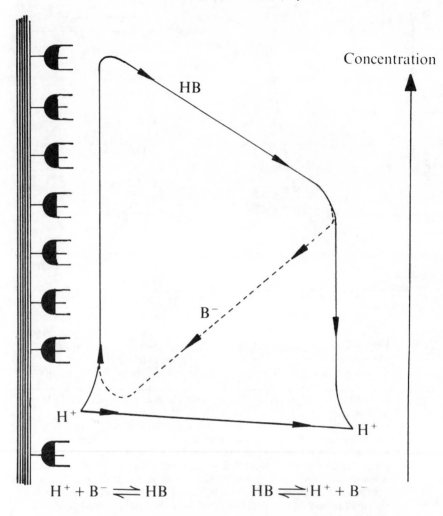

Figure 3.21. Diagram showing the facilitated transport of hydrogen ions away from an immobilised enzyme catalysing a reaction producing hydrogen ions. The hydrogen ion concentration gradient is small due to the low concentration of hydrogen ions in the bulk and the inability to produce a substantial hydrogen ion concentration (e.g. pH < 3) in the microenvironment. The buffer, represented by the conjugate pair HB/B$^-$, removes protons from the surface down much steeper concentration gradients, depending on its concentration and pK_a relative to the pH of the microenvironmental and bulk phases. The proton removal and delivery reactions occurring in these microenvironmental and bulk phases, respectively, are shown at the bottom.

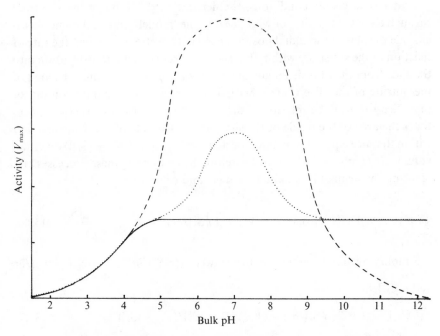

Figure 3.22. The effect of facilitated transport of hydrogen ions away from an immobilised enzyme catalysing a reaction producing hydrogen ions on the pH–activity profile (under conditions when μ is 10). ——— No buffer; –––––– high buffer concentration, $pK_a = 7$; ...... low buffer concentration, $pK_a = 7$. The buffer is only effective, at facilitating transport, close to its pK_a, as at low pH there is insufficient base present to bind the hydrogen ions and at high pH very little binding can occur due to the unfavourable equilibria.

purpose. The kinetics of other types of porous biocatalyst (e.g. flat membranous sheets, cylindrical pellets and fibres) may be analysed using a similar approach to that outlined here.

A porous spherical particle of immobilised enzyme may be represented as shown in Figure 3.6. The simplified model used in this analysis also requires that (1) the immobilised enzyme is uniformly distributed throughout the totally porous particle (i.e. ϵ and τ are both unity), (2) the Michaelis–Menten model describes the enzyme's kinetics, (3) the system is operating under steady-state conditions and is isothermal, (4) the diffusion of substrate and product obeys Fick's law (i.e. they are proportional to their concentration gradients), and the effective diffusivities are constant throughout the particle, (5) there is no external diffusional resistance (i.e. the substrate concentration at the particle surface ($[S_R]$) equals that in the bulk of the solution ($[S_o]$), and (6) neither partition nor inhibition occur.

Under steady-state conditions, the net rate of diffusion of the substrate through a concentric slice of width δr into the 'model' spherical immobilised enzyme particle at the radial position r from its centre must equal the rate of reaction of the substrate within that slice. The rate of substrate diffusion into the slice from the outside equals the flux $(D_S(\delta[S_r]/\delta r)_{r+\delta r})$ times the area of the outside of the slice $(4\pi(r + \delta r)^2)$. The rate of substrate diffusion out of that slice towards the interior of the particle equals the flux $(D_s(\delta[S_r]/\delta r)_r)$ times the area of the inside of the slice $(4\pi r^2)$. The rate of substrate reaction within the slice equals the volumetric activity $(V_{max}[S_r]/(K_m + [S_r]))$ times the volume $((4/3)\pi[(r + \delta r)^3 - r^3])$. Therefore, by combining these processes and ignoring the negligibly small terms in δr^2 and δr^3:

$$4\pi(r^2 + 2r\delta r)D_S(\delta[S_r]/\delta r)_{r+\delta r} - 4\pi r^2 D_s(\delta[S_r]/\delta r)_r = \frac{4\pi r^2 \delta r V_{max}[S_r]}{(K_m + [S_r])} \quad (3.46)$$

Simplifying and using the identity between $((\delta[S_r]/\delta r)_{r+\delta r} - (\delta[S_r]/\delta r)_r)/\delta r$ and $\delta^2[S_r]/\delta r^2$:

$$r^2(\delta^2[S_r]/\delta r^2) + 2r(\delta[S_r]/\delta r) = (V_{max}/D_s)r^2[S_r]/(K_m + [S_r]) \quad (3.47)$$

This may be further simplified by substitution with the dimensionless units; ρ for r/R, S for $[S_r]/[S_R]$, and β for $[S_R]/K_m$:

$$\rho^2(\delta^2 S/\delta\rho^2) + 2\rho(\delta S/\delta\rho) = (V_{max}R^2/D_S[S_R])\rho^2\beta S/(1 + \beta S) \quad (3.48)$$

The relative effects of internal diffusion and the kinetic rate of reaction can be described by use of a substrate modulus for internal diffusion (ϕ) which is conceptually similar to μ for external diffusion. ϕ, however, is defined differently for each type of porous biocatalyst (e.g. porous spheres, flat porous membranes, pellicular particles). Substituting ϕ for $(V_{max}R^2/D_s[S_R])$ and combining the left-hand terms in equation (3.48) gives:

$$(1/\rho^2)\delta(\rho^2(\delta S/\delta\rho))/\delta\rho = \phi\beta S/(1 + \beta S) \quad (3.49)$$

Therefore, as the slice width $\delta\rho$ tends to zero:

$$(1/\rho^2)d(\rho^2(dS/d\rho))/d\rho = \phi\beta S/(1 + \beta S) \quad (3.50)$$

An alternative definition of ϕ (ϕ^*) which is linear with respect to the characteristic length of the system (in this case this is the ratio of the volume to surface area $(R/3)$) is sometimes used, where:

$$\phi^* = (R/3)(V_{max}/(D_S K_m))^{1/2} \quad (3.51)$$

and, therefore, the two moduli are related by the expression:

$$\phi\beta = 9(\phi^*)^2 \quad (3.52)$$

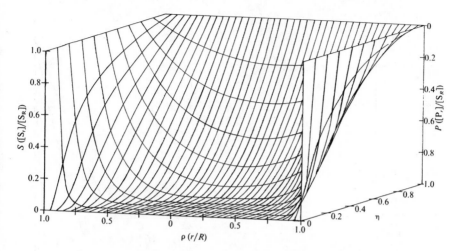

Figure 3.23. Substrate concentration profiles across spherical porous biocatalytic particles. The axes depict the substrate concentration relative to the external substrate concentration (S, which represents $[S_r]/[S_R]$), the dimensionless radial position (ρ, representing r/R) and the effectiveness factor (η). The profiles were derived by changing the substrate modulus (ϕ) whilst keeping β ($[S_R]/K_m$) constant (and equal to 0.1). Product concentration profiles are also shown; they are calculated assuming zero bulk product concentration and that the diffusion coefficients of substrate and product are equal.

This definition corresponds closer to that of μ, the substrate modulus for external diffusion, in that it is linear with respect to length. Thus, increases in the radius of porous particles are shown to be an additional factor which may cause diffusional control in an immobilised-enzyme catalysed reaction. Equation (3.50) cannot be solved analytically but may be solved by numerical methods using the boundary conditions that S and ρ are unity at the exterior surface of the particle, and ρ and $dS/d\rho$ are zero at the centre of the particle. This latter condition is necessary for reasons of symmetry through the centre of the particle. The solution is achieved by an iterative choice of the substrate concentration at the centre of the particle and using the relationship (3.49) to describe the changes in the substrate concentration, in small steps, from the centre outwards to the surface of the particle until the first boundary condition is met to within the accuracy required. The resultant concentration profile, which can be obtained rapidly using a fairly simple microcomputer, enables the overall rate and effectiveness of reaction, catalysed by the particle, to be calculated. Clearly, only if the substrate concentration is unchanged throughout the particle will the effectiveness of the enzyme be

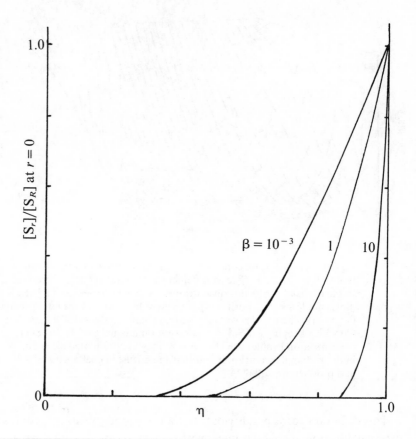

Figure 3.24. The variation of the substrate concentration in the centre of immobilised-enzyme particles with the effectiveness factor (η). Values for β (representing $[S_R]/K_m$) of 10, 1 and 10^{-3} are illustrated.

unchanged relative to freely soluble enzyme (i.e. $\eta = 1$). This is unlikely to be approached except in the case of very low enzyme loading or very small particles. The substrate modulus (ϕ) indicates the importance of these factors, being proportional to the enzyme loading and to the square of the particle diameter and inversely proportional to the diffusion coefficient. In this example, it is also inversely proportional to the bulk substrate concentration as this governs the extreme value obtainable by the concentration gradient.

The substrate concentration profiles across the particles at various effectiveness factors are shown in Figure 3.23. It can be seen that even at high

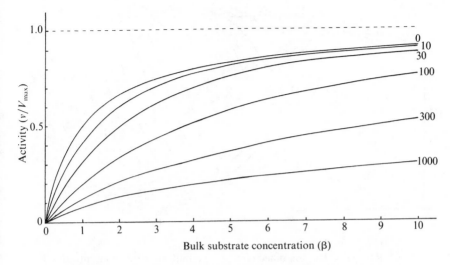

Figure 3.25. The variation in the rate of reaction catalysed by porous spherical particles containing immobilised enzyme with the dimensionless surface concentration of substrate, β (representing $[S_R]/K_m$). The top curve represents the case where there is no diffusional control (i.e. zero φ') whereas the lower curves show the effect of progressively greater normalised substrate modulus for internal diffusion, φ' (equalling 10, 30, 100, 300 and 1000). φ' is the substrate modulus for internal diffusion (φ) normalised with respect to the K_m (i.e. φ' = βφ) in order to make it independent of the absolute value of the exterior substrate concentration. The shapes of the curves displayed should be compared with those encountered under conditions of external diffusional control (Figure 3.12). Particularly note the absence here of the significant linear portion at low β and high substrate modulus.

η the substrate concentration drops significantly towards the centre of the particles and at low η this drop is so severe that the centres of the particles encounter very little substrate (Figure 3.24). This is particularly apparent at high β ($[S_R]/K_m$) as the reaction rate is close to V_{max} and small changes in the substrate concentration do not significantly lower the effectiveness factor. The variation in the rate of reaction with substrate concentration and substrate modulus is shown in Figure 3.25. At low effectiveness factors, only the outer layer of the biocatalytic particle is utilised, causing the particles to show an impressive apparent stability with time; relatively unused enzyme within the particle core is only brought into use as the surface immobilised enzyme inactivates due to denaturation.

The variation of the effectiveness factor with the substrate modulus and the dimensionleses substrate concentration is shown in Figure 3.26, which should be compared with the equivalent relationship for external diffusion

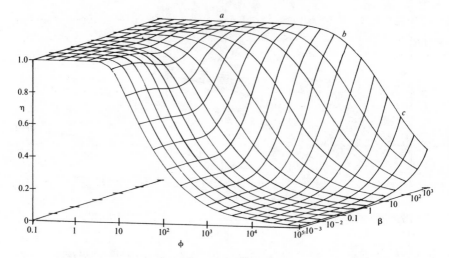

Figure 3.26. The combined effect of the bulk substrate concentration β (representing $[S_R]/K_m$) and substrate modulus for internal diffusion (φ) on the effectiveness factor (η) of porous spherical biocatalytic particles. The plateau (*a*) is an area of kinetic control, the graphical surface dropping through an area of intermediate control (*b*) to an area of diffusional control (*c*).

shown in Figure 3.13. From this it can be seen that values of the substrate modulus below unity have little effect on the productivity of the immobilised-enzyme particles, but higher values result in a considerable reduction in the effectiveness of the enzyme, especially at low substrate concentrations.

In real systems, the effectiveness factor is further reduced by steric effects which have been ignored in the above analysis. Where the substrate is large compared with the pore diameter, the effective diffusivity of the substrate within the pores will be significantly reduced, increasing φ. This reduction is generally proportional to the tortuosity of the pore geometry (τ) and inversely proportional to the particle porosity (ε; see legend to Figure 3.6 for definitions of τ and ε). The effective diffusion coefficient is additionally reduced as the ratio of the effective diameter of the substrate increases relative to the pore diameter, particularly where this ratio exceeds 0.02. This relationship usually causes a decrease in the effective diffusivity with the depth penetrated by the substrate. The active site of the enzyme may additionally be masked from binding the substrate by the difficulty with which the substrate can rotate within the confined space of the pores to give the correct effective conformation. The effectiveness factor may be increased by non-isothermal conditions where the reaction generates heat within the particles which is then unable to escape rapidly to the bulk phase. Non-isothermal operation is only rarely encountered, as most enzyme-catalysed reactions generate little heat and the catalytic particles are fairly small, having, for their volume, large surface areas through which the heat may escape.

Under circumstances where both external and internal diffusion gradients are found, the flux of substrate through the stagnant layer ($\{-D_S(d[S]/dr)\}_{external}$) must equal the flux entering the surface of the particles ($\{-D_s(d[S]/dr)\}_{internal}$):

$$\{-D_S(d[S]/dr)\}_{external} = \{-D_s(d[S]/dr)\}_{internal} \qquad (3.53)$$

As the effective diffusion coefficient of the substrate within the particles (D_s) is usually less than that in free solution, the substrate concentration gradient ($d[S]/dr$) within the particles must be greater than that occurring outside.

Determination of the intrinsic kinetic constant (K_m) is more complex in the case of internal than external diffusional control, due to the added complexity concerning variation in the effective substrate diffusivity. It is best determined under conditions where ϕ is less than unity, when little diffusional effect is apparent. Such conditions can be achieved by use of sufficiently small particles or low enzyme loading. Knowledge of the intrinsic K_m value(s) and V_{max} (obtained at high substrate concentrations, see Figure 3.27) allows the effective diffusion coefficients to be calculated.

The effect of internal diffusional control on reversible or inhibited reactions is similar to that encountered under external diffusional control. It may be analysed as outlined earlier for uninhibited non-reversible reactions, but replacing the equation for the volumetric activity by one involving reversibility or inhibition. For example, in the case of the reversible glucose isomerase reaction, the Michaelis–Menten volumetric activity term in equation (3.46) may be replaced by that from equation (1.58) (p. 25). This gives the following relationship, after a similar derivation to that shown in equations (3.46) to (3.49):

$$(1/\rho^2)\delta(\rho^2(\delta S/\delta\rho))/\delta\rho = \phi\beta(S - P/K_{eq})/(1 + \beta S + K_{SP}\beta P) \qquad (3.54)$$

where P represents $[P_r]/[S_R]$, $[P_r]$ is the product concentration at radial position r and K_{SP} replaces the Km^S/Km^P. Under steady-state conditions, the diffusion of substrate inwards and that of product outwards are linked by the relationship:

$$D_S([S_R] - [S_r]) = D_P([P_r] - [P_R]) \qquad (3.55)$$

where D_P is the diffusion coefficient of the product.
Therefore:

$$P = [P_R]/[S_R] + (D_S/D_P)(1 - S) \qquad (3.56)$$

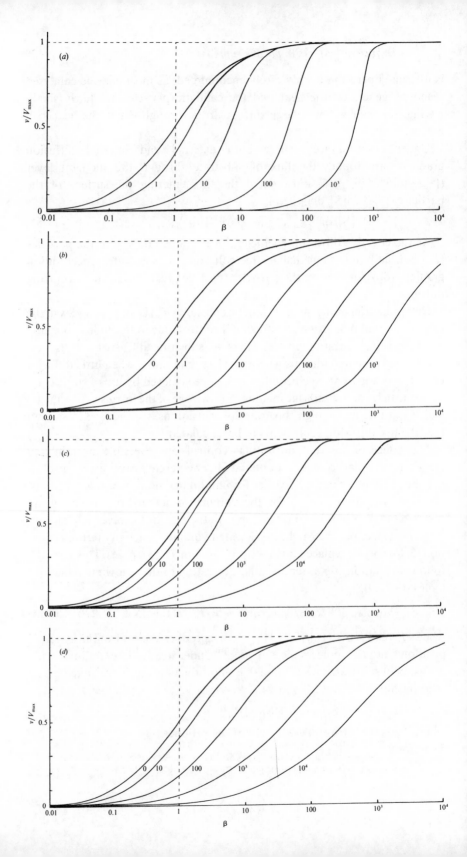

This value for P may be substituted into equation (3.54) and the resultant equation solved by an iterative numerical method as outlined previously.

A comparison of the rates of reaction for non-reversible and reversible reactions under both internal and external diffusion control are shown in Figures 3.27 and 3.28. Figure 3.27 emphasises the fact that diffusional control of a reaction can be overcome at sufficiently high substrate concentrations whatever the type of control or substrate modulus (so long as the substrate remains soluble and no substrate inhibition occurs). The two types of diffusional control may be distinguished at the higher values of substrate modulus by the pronounced differences in the steepness of the sigmoidal curve in the intermediary range of reaction rates (e.g. between 10% and 90% of V_{max}). This is because the rate of internally diffused-controlled reactions at low substrate concentrations are increased somewhat by the higher substrate flux through the outer layers of the porous biocatalysts. This effect can also be seen in the Eadie–Hofstee plots (Figure 3.28) at low rates of reaction. In practice, these graphs suffer from drawbacks if they are to be used to distinguish internal from external diffusional control. The semi-logarithmic plots need the possibility of high substrate concentrations without these adversely affecting the reaction, whereas the Eadie–Hofstee plots are most prone to error within the area of interest.

The difference between internal and external diffusional control is most noticeable in the variation in the rate of reaction with temperature. Reactions catalysed under conditions of external diffusional control obey equation (3.27). Their rates are independent of the activity of the enzyme and are, therefore, also almost independent of the temperature. Clearly, violent changes in temperature may so affect the enzyme that the rate of reaction is reduced below the rate at which the substrate can diffuse from the bulk of the solution, but then the reaction is no longer diffusionally controlled. Reactions catalysed under conditions of internal diffusional control do not obey equation (3.27). Increasing temperatures increase the rate at which the immobilised enzyme catalyses the reaction. This increases the substrate concentration gradient causing an increase in the flux of the substrate through the outer layer of the biocatalytic particles. Effectively, this halves the standard free energy of activation for the reaction relative to that

Figure 3.27. Semi-logarithmic plots for externally ((a) and (b); $\mu = 0$, 1, 10, 100 and 1000) and internally ((c) and (d)); $\phi' = \beta\phi = 0$, 10, 100, 1000, and 10 000) controlled reactions involving immobilised biocatalysts, (a) and (c) Non-reversible reactions; (b) and (d) have been calculated for the reversible reaction involving glucose isomerase ($K_{eq} = 1.14$, bulk concentration of glucose (S) and fructose (P) being 51% and 42% (w/w, respectively, of the total carbohydrate).

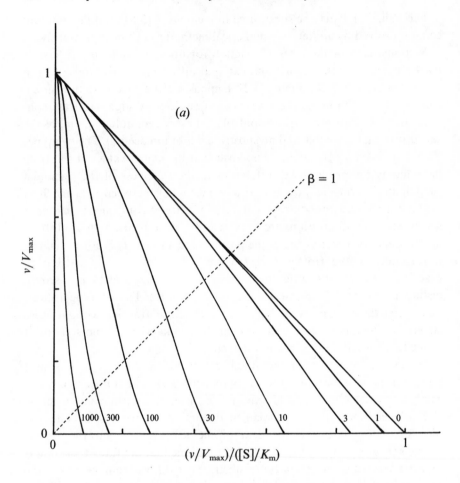

catalysed by the free enzyme in solution. It should be noted that any such reduction in activation energy will reduce the effect of increased temperature on the reaction rates of immobilised enzymes. The relationship between the rate of reaction and temperature is shown schematically in Figure 3.29.

Enzymic depolymerisation (including hydrolysis) of macromolecules may be affected by diffusional control. Large molecules diffuse only fairly slowly. After reaction, catalysed by the biocatalyst, the cleaved fragments normally retain their ability to act as substrates for the enzyme. They may diffuse away but are likely to be cleaved several times whilst in the vicinity of the immobilised enzyme. This causes a significant difference in the molecular weight profiles of the fragments produced by the use of free and immobilised enzymes. After a small degree of hydrolysis, most substrate molecules are cleaved by free soluble enzyme, whereas immobilised enzyme produces

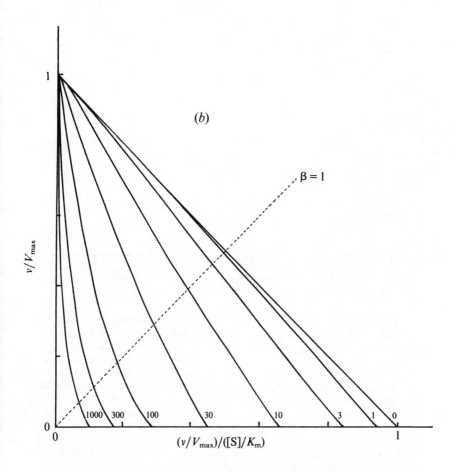

Figure 3.28. Eadie–Hofstee plots for internally ($\phi' = \beta\phi = 0$, 1, 3, 10, 30, 100, 300 and 1000) controlled reactions involving immobilised biocatalysts. (*a*) Non-reversible reaction, (*b*) reversible reaction involving glucose isomerase ($K_{eq} = 1.14$, bulk concentration of glucose (S) and fructose (P) being 51% and 42% (w/w), respectively, of the total carbohydrate). These plots correspond to those shown in Figure 3.27 (c) and (d). Figure 3.18 (a) and (b) shows equivalent plots for Figure 3.27 (a) and (b).

a small quantity of well-hydrolysed low molecular weight product, with the majority of the substrate molecules unchanged. This process is exacerbated by the use of porous biocatalysts, where there is some further restriction to the internal diffusion of large molecules. Use of immobilised enzyme is, therefore, indicated under circumstances where only a minimal proportion of partially hydrolysed product is required, as it generally

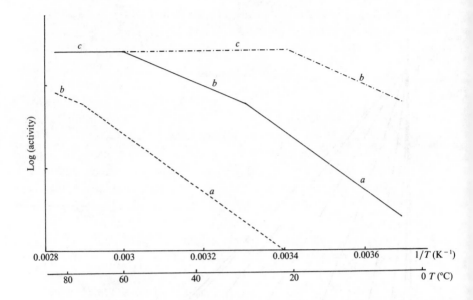

Figure 3.29. Schematic Arrhenius plots showing the progressive effect of diffusional limitations on the rate of reaction catalysed by porous particles containing immobilised enzyme. As the temperature increases, the activity progresses through three phases *a*, *b* and *c*, in that order. Curves *a* represent enzyme kinetic control of the reaction. The rate of reaction is sufficiently slow that no diffusional limitations are noticeable. The standard free energy of activation may be obtained from the gradient of this line. Curves *b* represent control of the reaction rate by the internal diffusion of the substrate (i.e. the intrinsic rate of reaction has increased to greater than the rate at which substrate can diffuse into the particles). Curves *c* represent control of the reaction rate by the diffusion of the substrate to the surface (i.e. the intrinsic rate of reaction has increased to greater than the rate at which substrate can diffuse through the stagnant layer surrounding the particles). No substrate is available for penetration into the pores. $------$ Low enzyme loading, $———$ intermediate enzyme loading, $-\cdot-\cdot-\cdot$ high enzyme loading. An Arrhenius plot using real data for an immobilised enzyme would show pronounced curvature between the three phases. The transition between the linear sections would not be readily discernible over the range of temperatures normally encountered for the use of immobilised enzymes, unless the standard free energy of activation for the reaction was unusually high (e.g. above 75000 J mol^{-1} for one of the transitions or above 100000 J mol^{-1} for both transitions).

produces a mixture of almost fully hydrolysed moieties and unchanged polymer.

The increase in the product concentration within the microenvironment may result in an increase in by-products caused by side-reactions, especially where the reverse reaction catalysed by enzyme does not show complete specificity for the re-formation of the substrates. This is particularly apparent in the action of some carbohydrases. The increased microenvironmental product concentration may be utilised, however, where a reaction pathway is required. Co-immobilisation of the necessary enzymes for the pathway results in a rapid conversion through the pathway, due to the localised high concentrations of the intermediates (Figure 3.30).

A logical extension of the use of immobilised multi-enzyme systems is the use of immobilised cell systems. These may be in a form which still allows respiration and reproduction or, in a restricted form, which cannot manage these functions but does retain catalytic activity. Sometimes they are treated with inhibitors or by physical means (e.g. heat) to ensure that only a subset of their natural enzymes remain active. Compared with immobilised multi-enzyme systems, immobilised cells are generally cheaper, easier to prepare with high activity, show little change in performance, on immobilisation, with respect to pH, ionic strength and temperature, and are generally preferred for use with metabolic pathways involving intracellular enzymes or dissociable cofactors or coenzymes. However, they do suffer from a number of practical disadvantages. They are more prone to microbial contamination, less efficient with respect to substrate conversion to product, much more difficult to control, and present diffusional problems due to their cell membranes.

Summary

(*a*) Immobilisation of enzymes enables their efficient and continuous use. The rationale behind immobilisation is the easy separation of product from the biocatalyst.

(*b*) Enzymes may be immobilised by adsorption, covalent binding, entrapment and membrane confinement, each method having its pros and cons. Adsorption is quick, simple and cheap but may be reversible. Covalent binding is permanent but expensive. Entrapment is generally applicable but may cause diffusional problems. Membrane confinement is a flexible method but expensive to set up.

(*c*) Immobilisation of enzymes may have a considerable effect on their kinetics. This may be due to structural changes to the enzyme and the creation of a distinct miocroenvironment around the enzyme. The activity of an immobilised enzyme is governed by the physical conditions within

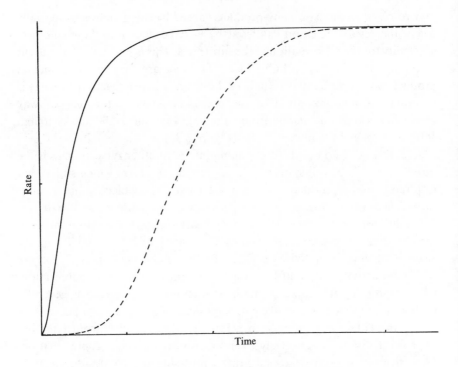

Figure 3.30. Diagram showing the effect of co-immobilisation on the rate of production through a short reaction pathway. ------ Mixture of free enzymes, in solution; ——— co-immobilised enzymes showing a much reduced lag phase. The reduction in the apparent lag phase is most noticeable when there are more enzymes in the pathway. It is least pronounced where the flux through the pathway is controlled by the first step as the microenvironmental concentration of the initial substrate cannot be higher than its bulk concentration but those of all intermediates may be raised due to diffusional restriction on their rate of efflux.

this microenvironment not those prevalent in the bulk phase. The immobilisation matrix affects the partition of material between the product phase and the enzyme phase and imposes restrictions on the rate of diffusion of material.

(*d*) Some effects of enzyme immobilisation are seen to be beneficial whilst others are detrimental to the economics of their use.

Bibliography

Bodálo, A., Gómez, J. L., Gómez, E., Bastida, J., Iborra, J. L. & Manjón, A. (1986). Analysis of diffusion effects on immobilized enzymes on porous supports with reversible Michaelis–Menten kinetics. *Enzyme and Microbial Technology*, **8**, 433–8.

Engasser, J.-M. & Coulet, P. R. (1977). Comparison of intrinsic stabilities of free and bound enzymes by graphical removal of diffusional effects. *Biochimica et Biophysica Acta*, **485**, 29–36.

Engasser, J.-M. & Horvath, C. (1973). Effect of internal diffusion in heterogeneous enzyme systems: evaluation of true kinetic parameters and substrate diffusivity. *Journal of Theoretical Biology*, **42**, 137–55.

Engasser, J.-M. & Horvath, C. (1974). Buffer-facilitated proton transport pH profile of bound enzymes. *Biochimica et Biophysica Acta*, **358**, 178–92.

Engasser, J.-M. & Horvath, C. (1976). Diffusion and kinetics with immobilised enzymes. In *Applied biochemistry and bioengineering*, vol. 1 *immobilised enzyme principles*, ed. L. B. Wingard, E. Katchalski-Katzir & L. Goldstein, pp. 127–220. New York: Academic Press.

European Federation of Biotechnology (1983). Guidelines for the characterization of immobilised biocatalysts. *Enzyme and Microbial Technology*, **5**, 304–7.

Goldstein, L. (1972). Microenvironmental effects on enzyme catalysis. A kinetic study of polyanionic and polycationic derivatives of chymotrypsin. *Biochemistry*, **11**, 4072–84.

Israelachvili, J. & Pashley, R. (1982). The hydrophobic interaction is long range, decaying exponentially with distance. *Nature*, **300**, 341–2.

Kennedy, J. F. & Cabral, J.M.S. (1987). Enzyme immobilisation. In *Biotechnology*, vol. 7a *Enzyme technology*, ed. J. F. Kennedy, pp. 347–404. Weinheim: VCH Verlagsgesellschaft mbH.

Martinek, K., Klibanov, A. M., Goldmacher, V. S. & Berezin, I. V. (1977*a*). The principles of enzyme stabilization. 1. Increase in thermostability of enzymes covalently bound to a complementary surface of a polymer support in a multipoint fashion. *Biochimica et Biophysica Acta*, **485**, 1–12.

Martinek, K., Klibanov, A. M., Goldmacher, V. S., Tchernysheva, A. V., Mozhaev, V. V., Berezin, I. V. & Glotov, B. O. (1977*b*). The principles of enzyme stabilization. 2. Increase in the thermostability of enzymes as a result of multipoint noncovalent interaction with a polymeric support. *Biochimica et Biophysica Acta*, **485**, 13–28.

Woodward, J. (1985). Immobilised enzymes: adsorption and covalent coupling. In *Immobilised cells and enzymes: a practical approach*, ed. J. Woodward, pp. 3–17. Oxford: IRL Press Ltd.

4   The large-scale use of enzymes in solution

Many of the more important industrially useful enzymes have been referred to earlier (see Table 2.1, pp. 42–3). The value of the world enzyme market has rapidly increased recently from £110M in 1960, £200M in 1970, £270M in 1980, £500M in 1985 to an estimated £1000M for 1990, representing an increase from 10% of the total catalyst market in 1980 to almost 20% in the 1990s. This increase has reflected the rise in the number of enzymes available on an industrial scale at relatively decreasing cost and the increasing wealth of knowledge concerning enzymes and their potential applications. As enzyme costs generally represent a small percentage at most, of the cost of the final product, it can be deduced that enzymes are currently involved in industrial processes with annual turnovers totalling many billions of pounds. Several enzymes, especially those used in starch processing, high-fructose syrup manufacture, textile desizing and detergent formulation, are now traded as commodity products on the world's markets. Although the cost of enzymes for use at the research scale is often very high, where there is a clear large-scale need for an enzyme its relative cost reduces dramatically with increased production.

Relatively few enzymes, notably those in detergents, meat tenderisers and garden composting agents, are sold directly to the public. Most are used by industry to produce improved or novel products, to bypass long and involved chemical synthetic pathways or for use in the separation and purification of isomeric mixtures. Many of the most useful, but least-understood, uses of free enzymes are in the food industry. Here they are used, together with endogenous enzymes, to produce or process foodstuffs, which are only rarely substantially refined. Their action, however apparently straightforward, is complicated due to the effect that small amounts of by-products or associated reaction products have on such subjective effects as taste, smell, colour and texture.

The use of enzymes in the non-food (chemicals and pharmaceuticals) sector is relatively straightforward. Products are generally separated and purified and, therefore, they are not prone to the subtleties available to food products. Most such enzymic conversions benefit from the use of immobilised enzymes or biphasic systems and will be considered in detail in Chapters 5 and 7.

The use of enzymes in detergents

The use of enzymes in detergent formulations is now common in developed countries, with over half of all detergents presently available containing enzymes. In spite of the fact that the detergent industry is the largest single market for enzymes at 25–30% of total sales, details of the enzymes used and the ways in which they are used, have rarely been published.

Dirt comes in many forms and includes proteins, starches and lipids. In addition, clothes that have been starched must be freed of the starch. Using detergents in water at high temperatures and with vigorous mixing, it is possible to remove most types of dirt but the cost of heating the water is high and lengthy mixing or beating will shorten the life of clothing and other materials. The use of enzymes allows lower temperatures to be employed and shorter periods of agitation are needed, often after a preliminary period of soaking. In general, enzyme detergents remove protein from clothes soiled with blood, milk, sweat, grass, etc. far more effectively than non-enzyme detergents. However, using modern bleaching and brightening agents, the difference between looking clean and being clean may be difficult to discern. At present only proteases and amylases are commonly used. Although a wide range of lipases is known, it is only very recently that lipases suitable for use in detergent preparations have been described.

Detergent enzymes must be cost-effective and safe to use. Early attempts to use proteases foundered because of producers and users developing hyper-sensitivity. This was combatted by developing dust-free granulates (about 0.5 mm in diameter) in which the enzyme is incorporated into an inner core, containing inorganic salts (e.g. NaCl) and sugars as preservative, bound with reinforcing fibres of carboxymethyl cellulose or similar protective colloid. This core is coated with inert waxy materials made from paraffin oil or polyethylene glycol plus various hydrophilic binders, which later disperse in the wash. This combination of materials both prevents dust formation and protects the enzymes against damage by other detergent components during storage.

Enzymes are used in surprisingly small amounts in most detergent prepar-ations, only 0.4–0.8% crude enzyme by weight (about 1% by cost). It follows that the ability to withstand the conditions of use is a more important criterion than extreme cheapness. Once released from its granulated form the enzyme must withstand anionic and non-ionic detergents, soaps, oxidants such as sodium perborate which generate hydrogen peroxide, optical

Table 4.1   *Compositions of an enzyme detergent*

Constituent	Composition (%)
Sodium tripolyphosphate (water softener, loosens dirt)[a]	38.0
Sodium alkane sulphonate (surfactant)	25.0
Sodium perborate tetrahydrate (oxidising agent)	25.0
Soap (sodium alkane carboxylates)	3.0
Sodium sulphate (filler, water softener)	2.5
Sodium carboxymethyl cellulose (dirt-suspending agent)	1.6
Sodium metasilicate (binder, loosens dirt)	1.0
Bacillus protease (3% active)	0.8
Fluorescent brighteners	0.3
Foam-controlling agents	Trace
Perfume	Trace
Water	to 100%

[a] A recent trend is to reduce this phosphate content for environmental reasons. It may be replaced by sodium carbonate plus extra protease.

brighteners and various less-reactive materials (Table 4.1), all at pH values between 8.0 and 10.5. Although one effect of incorporating enzymes is that lower washing temperatures may be employed with consequent savings in energy consumption, the enzymes must retain activity up to 60 °C.

The enzymes used are all produced using species of *Bacillus*, mainly by just two companies. Novo Industri A/S produce and supply three proteases, Alcalase, from *B. licheniformis*, Esperase, from an alkalophilic strain of a *B. licheniformis* and Savinase, from an alkalophilic strain of *B. amyloliquefaciens* (often mistakenly attributed to *B. subtilis*). Gist-Brocades produce and supply Maxatase, from *B. licheniformis*. Alcalase and Maxatase (both mainly subtilisin) are recommended for use at 10–65 °C and pH 7–10.5. Savinase and Esperase may be used at up to pH 11 and 12, respectively. The α-amylase supplied for detergent use is Termamyl, the enzyme from *B. licheniformis* which is also used in the production of glucose syrups. α-Amylase is particularly useful in dish-washing and de-starching detergents.

In addition to the granulated forms intended for use in detergent powders, liquid preparations in solution in water and slurries of the enzyme in a non-ionic surfactant are available for formulating in liquid 'spotting' concentrates, used for removing stubborn stains. Preparations containing both Termamyl and Alcalase are produced, Termamyl being sufficiently resistant to proteolysis to retain activity for long enough to fulfil its function.

It should be noted that all the proteolytic enzymes described are fairly non-specific serine endoproteases, giving preferred cleavage on the carboxyl side of hydrophobic amino acid residues but capable of hydrolysing most peptide links. They convert their substrates into small, readily soluble fragments which can be removed easily from fabrics. Only serine proteases may be used in detergent formulations: thiol proteases (e.g. papain) would be oxidised by the bleaching agents, and metalloproteases (e.g. thermolysin) would lose their metal cofactors due to complexing with the water softening agents or hydroxyl ions.

The enzymes are supplied in forms (as described above) suitable for formulation by detergent manufacturers. Domestic users are familiar with powdered preparations but liquid preparations for home use are increasingly available. Household laundering presents problems quite different from those of industrial laundering: the household wash consists of a great variety of fabrics soiled with a range of materials and the user requires convenience and effectiveness with less consideration of the cost. Home detergents will probably include both an amylase and a protease and a lengthy warm-water soaking time will be recommended. Industrial laundering requires effectiveness at minimum cost so heated water will be re-used if possible. Large laundries can separate their 'wash' into categories and thus minimise the usage of water and maximise the effectiveness of the detergents. Thus white cotton uniforms from an abattoir can be segregated for washing, only protease being required. A pre-wash soaking for 10–20 min at pH up to 11 and 30–40 °C is followed by a main wash for 10–20 min at pH 11 and 60–65 °C. The water from these stages is discarded to the sewer. A third wash includes hypochlorite as bleach which would inactivate the enzymes rapidly. The water from this stage is used again for the pre-wash but, by then, the hypochlorite concentration is insufficient to harm the enzyme. This is essentially a batch process: hospital laundries may employ continuous washing machines, which transfer less-initially-dirty linen from a pre-rinse initial stage, at 32 °C and pH 8.5, into the first wash at 60 °C and pH 11, then to a second wash, containing hydrogen peroxide, at 71 °C and pH 11, then to a bleaching stage and rinsing. Apart from the pre-soak stage, from which water is run to waste, the process operates counter-currently. Enzymes are used in the pre-wash and in the first wash, the levels of peroxide at this stage being insufficient to inactivate the enzymes.

There are opportunities to extend the use of enzymes in detergents both geographically and numerically. They have not found widespread use in developing countries which are often hot and dusty, making frequent washing of clothes necessary. The recent availability of a suitable lipase may increase the quantities of enzymes employed very significantly. There are, perhaps, opportunities for enzymes such as glucose oxidase, lipoxygenase and glycerol oxidase as means of generating hydrogen peroxide *in situ*. Added peroxidases may aid the bleaching efficacy of this peroxide.

A recent development in detergent enzymes has been the introduction of an alkaline-stable fungal cellulase preparation for use in washing cotton fabrics. During use, small fibres are raised from the surface of cotton thread, resulting in a change in the 'feel' of the fabric and, particularly, in the lowering of the brightness of colours. Treatment with cellulase removes the small fibres without apparently damaging the major fibres and restores the fabric to its 'as new' condition. The cellulase also aids the removal of soil particles from the wash by hydrolysing associated cellulose fibres.

Applications of proteases in the food industry

Certain proteases have been used in food processing for centuries and any record of the discovery of their activity has been lost in the mists of time. Rennet (mainly chymosin), obtained from the fourth stomach (abomasum) of unweaned calves has been used traditionally in the production of cheese. Similarly, papain from the leaves and unripe fruit of the pawpaw (*Carica papaya*) has been used to tenderise meats. These ancient discoveries have led to the development of various food applications for a wide range of available proteases from many sources, usually microbial. Proteases may be used at various pH values, and they may be highly specific in their choice of cleavable peptide links or quite non-specific. Proteolysis generally increases the solubility of proteins at their isoelectric points.

The action of rennet in cheese making is an example of the hydrolysis of a specific peptide linkage, between phenylalanine and methionine residues (-Phe105-Met106-) in the κ-casein protein present in milk (see reaction scheme [1.3], p. 3). The κ-casein acts by stabilising the colloidal nature of the milk, its hydrophobic N-terminal region associating with the lipophilic regions of the otherwise insoluble α- and β-casein molecules, whilst its negatively charged C-terminal region associates with the water and prevents the casein micelles from growing too large. Hydrolysis of the labile peptide linkage between these two domains, resulting in the release of a hydrophilic glycosylated and phosphorylated oligopeptide (caseino macropeptide) and the hydrophobic *para*-κ-casein, removes this protective effect, allowing coagulation of the milk to form curds, which are then compressed and turned

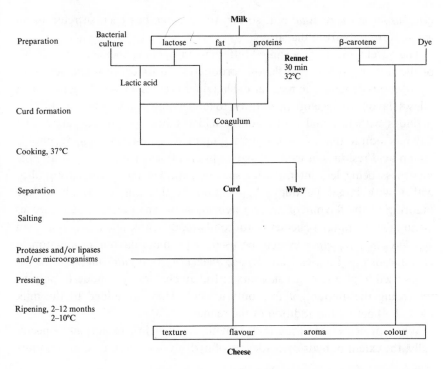

Figure 4.1. Outline method for the preparation of cheese.

into cheese (Figure 4.1). The coagulation process depends upon the presence of Ca^{2+} and is very temperature dependent ($Q_{10} = 11$) and so can be controlled easily. Calf rennet, consisting of mainly chymosin with a small but variable proportion of pepsin, is a relatively expensive enzyme and various attempts have been made to find cheaper alternatives from microbial sources. These have ultimately proved to be successful and microbial rennets are used for about 70% of US cheese and 33% of cheese production world-wide.

The major problem that had to be overcome in the development of the microbial rennets was temperature lability. Chymosin is a relatively unstable enzyme and once it has done its major job, little activity remains. However, the enzyme from *Mucor miehei* retains activity during the maturation stages of cheese-making and produces bitter off-flavours. Treatment of the enzyme with oxidising agents (e.g. H_2O_2, peracids), which convert methionine residues to their sulphoxides, reduces its thermostability by about 10 °C and renders it more comparable with calf rennet. This is a rare example of enzyme technology being used to destabilise an enzyme. Attempts have been made to clone chymosin into *Escherichia coli* and

Saccharomyces cerevisiae but, so far, the enzyme has been secreted in an active form only from the latter.

The development of unwanted bitterness in ripening cheese is an example of the role of proteases in flavour production in foodstuffs. The action of endogenous proteases in meat after slaughter is complex but 'hanging' meat allows flavour to develop, in addition to tenderising it. It has been found that peptides with terminal acidic amino acid residues give meaty, appetising flavours akin to that of monosodium glutamate. Non-terminal hydrophobic amino acid residues in medium-sized oligopeptides give bitter flavours, the bitterness being less intense with smaller peptides and disappearing alto-gether with larger peptides. Application of this knowledge allows the tailoring of the flavour of protein hydrolysates. The presence of proteases during the ripening of cheeses is not totally undesirable and a protease from *Bacillus amyloliquefaciens* may be used to promote flavour production in cheddar cheese. Lipases from *Mucor miehei* or *Aspergillus niger* are some-times used to give stronger flavours in Italian cheeses by a modest lipolysis, increasing the amount of free butyric acid. They are added to the milk (30 U l^{-1}) before the addition of the rennet.

When proteases are used to depolymerise proteins, usually non-specifi-cally, the extent of hydrolysis (degree of hydrolysis) is described in *DH* units where:

$$DH = 100 \times \left(\frac{\text{Number of peptide bonds cleaved}}{\text{Initial number of peptide bonds present}} \right) \qquad (4.1)$$

Commercially, using enzymes such as subtilisin, *DH* values of up to 30 are produced using protein preparations of 8–12% (w/w). The enzymes are formulated so that the value of the enzyme : substrate ratio used is 2–4% (w/w). At the high pH needed for effective use of subtilisin, protons are released during the proteolysis and must be neutralised:

$$\text{subtilisin (pH 8.5)}$$
$$H_2N\text{-aa-aa-aa-aa-aa-COO}^- \xrightarrow{\hspace{2cm}} H_2N\text{-aa-aa-aa-COO}^-$$
$$+ H_2N\text{-aa-aa-COO}^- + H^+ \quad [\mathbf{4.1}]$$

where aa is an amino acid residue.

Correctly applied proteolysis of inexpensive materials such as soya protein can increase the range and value of their usage, as indeed occurs naturally in the production of soy sauce. Partial hydrolysis of soya protein, to around 3.5 *DH* greatly increases its 'whipping expansion', further hydrolysis, to around 6 *DH* improves its emulsifying capacity. If their flavours are correct, soya protein hydrolysates may be added to cured meats. Hydrolysed proteins may develop properties that contribute to the elusive, but valuable, phenomenon of 'mouth feel' in soft drinks.

Proteases are used to recover protein from parts of animals (and fish) that would otherwise go to waste after butchering. About 5% of the meat cannot be removed mechanically from bone. To recover this, bones are mashed and incubated at 60 °C with neutral or alkaline proteases for up to 4 h. The meat slurry produced is used in canned meats and soups.

Large quantities of blood are available but, except in products such as black puddings, they are not generally acceptable in foodstuffs because of their colour. The protein is of a high quality nutritionally and is de-haemed using subtilisin. Red cells are collected and haemolysed in water. Subtilisin is added and hydrolysis is allowed to proceed batchwise, with neutralisation of the released protons, to around 18 *DH*, when the hydrophobic haem molecules precipitate. Excessive degradation is avoided to prevent the formation of bitter peptides. The enzyme is inactivated by a brief heat treatment at 85 °C and the product is centrifuged; no residual activity is allowed into meat products. The haem-containing precipitate is recycled and the light-brown supernatant is processed through activated carbon beads to remove any residual haem. The purified hydrolysate, obtained in 60% yield, may be spray-dried and is used in cured meats, sausages and luncheon meats.

Meat tenderisation by the endogenous proteases in the muscle after slaughter is a complex process which varies with the nutritional, physiological and even psychological (i.e. frightened or not) state of the animal at the time of slaughter. Meat of older animals remains tough but can be tenderized by injecting inactive papain into the jugular vein of the live animal shortly before slaughter. Injection of the active enzyme would rapidly kill the animal in an unacceptably painful manner so the inactive oxidised disulphide form of the enzyme is used. On slaughter, the resultant reducing conditions cause free thiols to accumulate in the muscle, activating the papain and so tenderizing the meat. This is a very effective process as only 2–5 p.p.m of the inactive enzyme needs to be injected. Recently, however, it has found disfavour as it destroys the animals heart, liver and kidneys which otherwise could be sold and, being reasonably heat stable, its action is difficult to control and persists into the cooking process.

Proteases are also used in the baking industry. Where appropriate, dough may be prepared more quickly if its gluten is partially hydrolysed. A heat-labile fungal protease is used so that it is inactivated early in the subsequent baking. Weak-gluten flour is required for biscuits in order that the dough can be spread thinly and retain decorative impressions. In the past this has been obtained from European domestic wheat but this is being replaced by high-gluten varieties of wheat. The gluten in the flour derived from these must be extensively degraded if such flour is to be used efficiently for making biscuits or for preventing shrinkage of commercial pie pastry away from their aluminium dishes.

The use of proteases in the leather and wool industries

The leather industry consumes a significant proportion of the world's enzyme production. Alkaline proteases are used to remove hair from hides. This process is far safer and more pleasant than the traditional methods involving sodium sulphide. Relatively large amounts of enzyme are required (0.1–1.0 % (w/w)) and the process must be closely controlled to avoid reducing the quality of the leather. After dehairing, hides which are to be used for producing soft leather clothing and goods are bated, a process, often involving pancreatic enzymes, that increases their suppleness and improves the softness of their appearance.

Proteases have been used, in the past, to 'shrinkproof' wool. Wool fibres are covered in overlapping scales pointing towards the fibre tip. These give the fibres highly directional frictional properties, movement in the direction away from the tip being favoured relative to movement towards it. This propensity for movement solely in the one direction may lead to shrinkage and many methods have been used in attempts to eliminate the problem (e.g. chemical oxidation or coating the fibres in polymer). A successful method involved the partial hydrolysis of the scale tips with the protease papain. This method also gave the wool a silky lustre and added to its value. The method was abandoned some years ago, primarily for economic reasons. It is not unreasonable to expect its use to be re-established now that cheaper enzyme sources are available.

The use of enzymes in starch hydrolysis

Starch is the commonest storage carbohydrate in plants. It is used by the plants themselves, by microbes and by higher organisms so there is a great diversity of enzymes able to catalyse its hydrolysis. Starch from all plant sources occurs in the form of granules which differ markedly in size and physical characteristics from species to species. Chemical differences are less marked. The major difference is the ratio of amylose to amylopectin; e.g. corn starch from waxy maize contains only 2% amylose but that from amylomaize is about 80% amylose. Some starches, for instance from potato, contain covalently bound phosphate in small amounts (0.2% approximately), which has significant effects on the physical properties of the starch

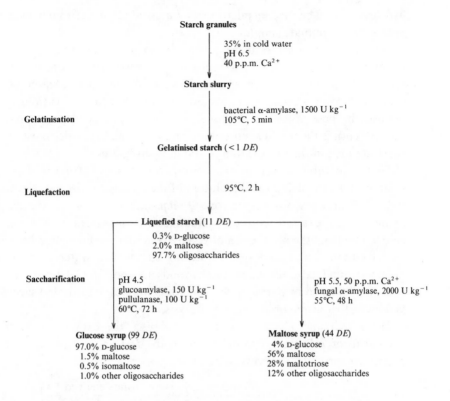

Figure 4.2. The use of enzymes in processing starch. Typical conditions are given.

but does not interfere with its hydrolysis. Acid hydrolysis of starch has had widespread use in the past. It is now largely replaced by enzymic processes, as it required the use of corrosion resistant materials, gave rise to high colour and saltash content (after neutralisation), needed more energy for heating and was relatively difficult to control.

Of the two components of starch, amylopectin presents the greater challenge to hydrolytic enzyme systems. This is due to the residues involved in α-1,6-glycosidic branch points which constitute about 4–6% of the glucose present. Most hydrolytic enzymes are specific for α-1,4-glucosidic links yet the α-1,6-glucosidic links must also be cleaved for complete hydrolysis of amylopectin to glucose. Some of the most impressive recent exercises in the development of new enzymes have concerned debranching enzymes.

It is necessary to hydrolyse starch in a wide variety of processes which may be condensed into two basic classes: (1) processes in which the starch hydrolysate is to be used by microbes or man, and (2) processes in which it is necessary to eliminate starch.

In the former processes, such as glucose syrup production, starch is usually the major component of reaction mixtures, whereas in the latter processes, such as the processing of sugar cane juice, small amounts of starch which contaminate non-starchy materials are removed. Enzymes of various types are used in these processes. Although starches from diverse plants may be utilised, corn is the world's most abundant source and provides most of the substrate used in the preparation of starch hydrolysates.

There are three stages in the conversion of starch (Figure 4.2): (1) *gelatinisation*, involving the dissolution of the nanogram-sized starch granules to form a viscous suspension; (2) *liquefaction*, involving the partial hydrolysis of the starch, with concomitant loss in viscosity; and (3) *saccharification*, involving the production of glucose and maltose by further hydrolysis. Gelatinisation is achieved by heating starch with water, and occurs necessarily and naturally when starchy foods are cooked. Gelatinised starch is readily liquefied by partial hydrolysis with enzymes or acids and saccharified by further acidic or enzymic hydrolysis.

The starch and glucose syrup industry uses the expression dextrose equivalent or *DE*, similar in definition to the *DH* units of proteolysis, to describe its products, where:

$$DE = 100 \times \left(\frac{\text{Number of glycosidic bonds cleaved}}{\text{Initial number of glycosidic bonds present}} \right) \qquad (4.2)$$

In practice, this is usually determined analytically by use of the closely related, but not identical, expression:

$$DE = 100 \times \left(\frac{\text{Reducing sugar, expressed as glucose}}{\text{Total carbohydrate}} \right) \qquad (4.3)$$

Thus, *DE* represents the percentage hydrolysis of the glycosidic linkages present. Pure glucose has a *DE* of 100, pure maltose has a *DE* of about 50 (depending upon the analytical methods used; see equation (4.3)) and starch has a *DE* of effectively zero. During starch hydrolysis, *DE* indicates the extent to which the starch has been cleaved. Acid hydrolysis of starch has long been used to produce 'glucose syrups' and even crystalline glucose (dextrose monohydrate). Very considerable amounts of 42 *DE* syrups are produced using acid and are used in many applications in confectionery. Further hydrolysis using acid is not satisfactory because of undesirably coloured and flavoured breakdown products. Acid hydrolysis appears to be a totally random process which is not influenced by the presence of α-1,6-glucosidic linkages.

Table 4.2 *Enzymes used in starch hydrolysis*

Enzyme	EC number	Source	Action
α-Amylase	3.2.1.1	*Bacillus amyloliquefaciens*	Only α-1,4-oligosaccharide links are cleaved to give α-dextrins and predominantly maltose (G_2), G_3, G_6 and G_7 oligosaccharides
		B. licheniformis	Only α-1,4-oligosaccharide links are cleaved to give α-dextrins and predominantly maltose, G_3, G_4 and G_5 oligosaccharides
		Aspergillus oryzae, A. niger	Only α-1,4 oligosaccharide links are cleaved to give α-dextrins and predominantly maltose and G_3 oligosaccharides
Saccharifying α-amylase	3.2.1.1	*B. subtilis* (*amylosacchariticus*)	Only α-1,4-oligosaccharide links are cleaved to give α-dextrins with maltose, G_3, G_4 and up to 50% (w/w) glucose
β-Amylase	3.2.1.2	Malted barley	Only α-1,4-links are cleaved, from non-reducing ends, to give limit dextrins and β-maltose
Glucoamylase	3.2.1.3	*A. niger*	α-1,4 and α-1,6-links are cleaved, from the non-reducing ends, to give β-glucose
Pullulanase	3.2.1.41	*B. acidopullulyticus*	Only α-1,6-links are cleaved to give straight-chain maltodextrins

The nomenclature of the enzymes used commercially for starch hydrolysis is somewhat confusing and the EC numbers sometimes lump together enzymes with subtly different activities (Table 4.2). For example, α-amylases may be subclassified as liquefying or saccharifying amylases but even this classification is inadequate to encompass all the enzymes that are used in commercial starch hydrolysis. One reason for the confusion in the nomenclature is the use of the anomeric form of the released reducing group in the product rather than that of the bond being hydrolysed; the products of bacterial and fungal α-amylases are in the α-configuration and the products of β-amylases are in the β-configuration, although all these enzymes cleave between α-1,4-linked glucose residues.

The α-amylases (1,4-α-D-glucan glucanohydrolases) are endohydrolases which cleave 1,4-α-D-glucosidic bonds and can bypass but cannot hydrolyse 1,6-α-D-glucosidic branchpoints. Commercial enzymes used for the industrial hydrolysis of starch are produced by *Bacillus amyloliquefaciens* (supplied by various manufacturers) and by *B. licheniformis* (supplied by Novo Industri A/S as Termamyl). They differ principally in their tolerance of high temperatures, Termamyl retaining more activity at up to 110 °C, in the presence of starch, than the *B. amyloliquefaciens* α-amylase. The maximum *DE* obtainable using bacterial α-amylases is around 40 but prolonged treatment leads to the formation of maltulose (4-α-D-glucopyranosyl-D-fructose), which is resistant to hydrolysis by glucoamylase and α-amylases. In most commercial processes where further saccharification is to occur *DE* values of 8–12 are used. The principal requirement for liquefaction to this extent is to reduce the viscosity of the gelatinised starch to ease subsequent processing.

Various manufacturers use different approaches to starch liquefaction using α-amylases but the principles are the same. Granular starch is slurried at 30–40% (w/w) with cold water, at pH 6.0–6.5, containing 20–80 p.p.m. Ca^{2+} (which stabilises and activates the enzyme) and the enzyme is added (via a metering pump). The α-amylase is usually supplied at high activities so that the enzyme dose is 0.5–0.6 kg tonne^{-1} (about 1500 U kg^{-1} dry matter) of starch. When Termamyl is used, the slurry of starch plus enzyme is pumped continuously through a jet cooker, which is heated to 105 °C using live steam. Gelatinisation occurs very rapidly and the enzymic activity, combined with the significant shear forces, begins the hydrolysis. The residence time in the jet cooker is very brief. The partly gelatinised starch is passed into a series of holding tubes maintained at 100–105 °C and held for 5 min to complete the gelatinisation process. Hydrolysis to the required *DE* is completed in holding tanks at 90–100 °C for 1 to 2 h. These tanks contain baffles to discourage backmixing. Similar processes may be used with *B. amyloliquefaciens* α-amylase but the maximum temperature of 95 °C must not be exceeded. This has the drawback that a final 'cooking' stage must be introduced when the required *DE* has been attained in order to gelatinise the recalcitrant starch grains present in some types of starch which would otherwise cause cloudiness in solutions of the final product.

The liquefied starch is usually saccharified but comparatively small amounts are spray-dried for sale as 'maltodextrins' to the food industry mainly for use as bulking agents and in baby food. In this case, residual enzymic activity may be destroyed by lowering the pH towards the end of the heating period.

Fungal α-amylase also finds use in the baking industry. It often needs to be added to bread-making flours to promote adequate gas production and starch modificiation during fermentation. This has become necessary since the introduction of combine harvesters. They reduce the time between cutting and threshing of the wheat, which previously was sufficient to allow a limited sprouting so increasing the amounts of endogenous enzymes. The fungal enzymes are used rather than those from bacteria as their action is easier to control due to their relative heat lability, denaturing rapidly during baking.

Production of glucose syrup

The liquefied starch at 8–12 *DE* is suitable for saccharification to produce syrups with *DE* values of from 45 to 98 or more. The greatest quantities produced are the syrups with *DE* values of about 97. At present these are produced using the exoamylase, glucan 1,4-α-glucosidase (1,4-α-D-glucan glucohydrolase, commonly called glucoamylase but also called amyloglucosidase and γ-amylase), which releases β-D-glucose from 1,4-α-, 1,6-α- and 1,3-α-linked glucans. In theory, carefully liquefied starch at 8–12 *DE* can be hydrolysed completely to produce a final glucoamylase reaction mixture with *DE* of 100 but, in practice, this can be achieved only at comparatively low substrate concentrations. The cost of concentrating the product by evaporation decrees that a substrate concentration of 30% is used. It follows that the maximum *DE* attainable is 96–98 with syrup composition 95–97% glucose, 1–2% maltose and 0.5–2% (w/w) isomaltose (α-D-glucopyranosyl-(1,6)-D-glucose). This material is used after concentration, directly for the production of high-fructose syrups or for the production of crystalline glucose.

Whereas liquefaction is usually a continuous process, saccharification is most often conducted as a batch process. The glucoamylase most often used is produced by *Aspergillus niger* strains. This has a pH optimum of 4.0–4.5 and operates most effectively at 60 °C, so liquefied starch must be cooled and its pH adjusted before addition of the glucoamylase. The cooling must be rapid, to avoid retrogradation (the formation of intractable insoluble aggregates of amylose; the process that gives rise to the skin on custard). Any remaining bacterial α-amylase will be inactivated when the pH is lowered; however, this may be replaced later by some acid-stable α-amylase which is normally present in the glucoamylase preparations. When conditions are correct the glucoamylase is added, usually at the dosage of 0.65–0.80 litre

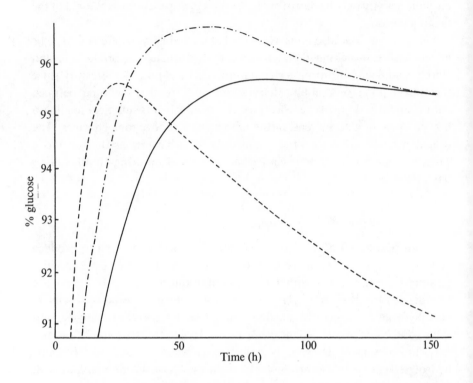

Figure 4.3. The % glucose formed from 30% (w/w) 12 *DE* maltodextrin, at 60 °C and pH 4.3, using various enzyme solutions. ———— 200 U kg^{-1d} *Aspergillus niger* glucoamylase; —————— 400 U kg^{-1} *A. niger* glucoamylase; —·—·—· 200 U kg^{-1} *A. niger* glucoamylase plus 200 U kg^{-1} *Bacillus acidopullulyticus* pullulanase. The relative improvement on the addition of pullulanase is even greater at higher substrate concentrations.

enzyme preparation tonne^{-1} starch (200 U kg^{-1}). Saccharification is normally conducted in vast stirred tanks, which may take several hours to fill (and empty), so time will be wasted if the enzyme is added only when the reactors are full. The alternatives are to meter the enzyme at a fixed ratio or to add the whole dose of enzyme at the commencement of the filling stage. The latter should give the most economical use of the enzyme.

The saccharification process takes about 72 h to complete but may, of course, be speeded up by the use of more enzyme. Continuous saccharification is possible and practicable if at least six tanks are used in series. It is necessary to stop the reaction, by heating to 85 °C for 5 min, when a maximum *DE* has been attained. Further incubation will result in a fall in the *DE*, to about 90 *DE* eventually, caused by the formation of isomaltose as accumulated glucose re-polymerises with the approach of thermodynamic equilibrium (Figure 4.3).

The saccharified syrup is filtered to remove fat and denatured protein released from the starch granules and may then be purified by passage through activated charcoal and ion-exchange resins. It should be remembered that the dry substance concentration increases by about 11% during saccharification, because one molecule of water is taken up for each glycosidic bond hydrolysed (molecule of glucose produced).

Although glucoamylase catalyses the hydrolysis of 1,6-α-linkages, their breakdown is slow compared with that of 1,4-α-linkages (e.g. the rates of hydrolysing the 1,4-α, 1,6-α and 1,3-α-links in tetrasaccharides are in the proportions 300 : 6 : 1). It is clear that the use of a debranching enzyme would speed the overall saccharification process but, for industrial use, such an enzyme must be compatible with glucoamylase. Two types of debranching enzymes are available: pullulanase, which acts as an exo-hydrolase on starch dextrins; and isoamylase (EC.3.2.1.68), which is a true endohydrolase. Novo Industri A/S have recently introduced a suitable pullulanase, produced by a strain of *Bacillus acidopullulyticus*. The pullulanase from *Klebsiella aerogenes* which has been available commercially for some time is unstable at temperatures over 45 °C but the *B. acidopullulyticus* enzymes can be used under the same conditions as the *Aspergillus* glucoamylase (60 °C, pH 4.0–4.5). The practical advantage of using pullulanase together with glucoamylase is that less glucoamylase need be used. This does not in itself give any cost advantage but because less glucoamylase is used and fewer branched oligosaccharides accumulate towards the end of the saccharification, the point at which isomaltose production becomes significant occurs at higher *DE* (Figure 4.3). It follows that higher *DE* values and glucose contents can be achieved when pullulanase is used (98–99 *DE* and 95–97% (w/w) glucose, rather than 97–98 *DE*) and higher substrate concentrations (30–40% dry solids rather than 25-30%) may be treated. The extra cost of using pullulanase is recouped by savings in evaporation and glucoamylase costs. In addition, when the product is to be used to manufacture high-fructose syrups, there is a saving in the cost of further processing.

The development of the *B. acidopullulyticus* pullulanase is an excellent example of what can be done if sufficient commercial pull exists for a new enzyme. The development of a suitable α-D-glucosidase, in order to reduce the reversion, would be an equally useful step for industrial glucose production. Screening of new strains of bacteria for a novel enzyme of this type is a major undertaking. It is not surprising that more details of the screening procedures used are not readily available.

Production of syrups containing maltose

Traditionally, syrups containing maltose as a major component have been produced by treating barley starch with barley β-amylase. β-Amylases (1,4-α-D-glucan maltohydrolases) are exohydrolases which release maltose from 1,4-α-linked glucans but neither bypass nor hydrolyse 1,6-α-linkages. *High-maltose syrups* (40–50 *DE*, 45–60% (w/w) maltose, 2–7% (w/w) glucose) tend not to crystallise, even below 0 °C and are relatively non-hygroscopic. They are used for the production of hard candy and frozen deserts. *High conversion syrups* (60–70 *DE*, 30–37% maltose, 35–43% glucose, 10% maltotriose, 15% other oligosaccharides, all by weight) resist crystallisation above 4 °C and are sweeter (Table 4.3). They are used for soft candy and in the baking, brewing and soft drinks industries. It might be expected that β-amylase would be used to produce maltose-rich syrups from corn starch, especially as the combined action of β-amylase and pullulanase give almost quantitative yields of maltose. This is not done on a significant scale nowadays because presently available β-amylases are relatively expensive, not sufficiently temperature stable (although some thermostable β-amylases from species of *Clostridium* have recently been reported) and are easily inhibited by copper and other heavy metal ions. Instead fungal α-amylases, characterised by their ability to hydrolyse maltotriose (G_3) rather than maltose (G_2), are employed often in combination with glucoamylase. Presently available enzymes, however, are not totally compatible; fungal α-amylases requiring a pH of not less than 5.0 and a reaction temperature not exceeding 55 °C.

High-maltose syrups (see Figure 4.2) are produced from liquefied starch of around 11 *DE* at a concentration of 35% dry solids using fungal α-amylase alone. Saccharification occurs over 48 h, by which time the fungal α-amylase has lost its activity. Now that a good pullulanase is available, it is possible to use this in combination with fungal α-amylases to produce syrups with even higher maltose contents.

High-conversion syrups are produced using combinations of fungal α-amylase and glucoamylase. These may be tailored to customers' specifications by adjusting the activities of the two enzymes used but inevitably, as glucoamylase is employed, the glucose content of the final product will be higher than that of high-maltose syrups. The stability of glucoamylase necessitates stopping the reaction, by heating, when the required composition is reached. It is now possible to produce starch hydrolysates with any *DE* between 1 and 100 and with virtually any composition using combinations of bacterial α-amylases, fungal α-amylases, glucoamylase and pullulanase.

Table 4.3   *The relative sweetness of food ingredients*

Food ingredient	Relative sweetness (by weight, solids)
Sucrose	1.0
Glucose	0.7
Fructose	1.3
Galactose	0.7
Maltose	0.3
Lactose	0.2
Raffinose	0.2
Hydrolysed sucrose	1.1
Hydrolysed lactose	0.7
Glucose syrup 11 *DE*	< 0.1
Glucose syrup 42 *DE*	0.3
Glucose syrup 97 *DE*	0.7
Maltose syrup 44 *DE*	0.3
High-conversion syrup 65 *DE*	0.5
HFCS (42% fructose)[a]	1.0
HFCS (55% fructose)	1.1
Aspartame	180

[a] HFCS, high-fructose corn syrup.

Enzymes in the sucrose industry

The sucrose industry is a comparatively minor user of enzymes but provides a few historically significant and instructive examples of enzyme technology. The hydrolysis ('inversion') of sucrose, completely or partially, to glucose and fructose provides sweet syrups that are more stable (i.e. less likely to crystallise) than pure sucrose syrups. The most familiar 'Golden Syrup' is produced by acid hydrolysis of one of the less pure streams from the cane sugar refinery but other types of syrup are produced using yeast (*Saccharomyces cerevisiae*) invertase. Although this enzyme is unusual in that it suffers from substrate inhibition at high sucrose levels (> 20% (w/w)), this does not prevent its commercial use at even higher concentrations:

$$\text{sucrose} \longrightarrow \alpha\text{-D-glucose} + \beta\text{-D-fructose} \qquad [4.2]$$

Traditionally, invertase was produced on site by autolysing yeast cells. The autolysate was added to the syrup (70% sucrose (w/w)) to be inverted together with small amounts of xylene to prevent microbial growth. Inversion was complete in 48–72 h at 50 °C and pH 4.5. The enzyme and xylene were removed during the subsequent refining and evaporation. Partially inverted syrups were (and still are) produced by blending totally inverted syrups with sucrose syrups. Now, commercially produced invertase concentrates are employed.

The production of hydrolysates of a low molecular weight compound in essentially pure solution seems an obvious opportunity for the use of an immobilised enzyme, yet this is not done on a significant scale, probably because of the extreme simplicity of using the enzyme in solution and the basic conservatism of the sugar industry.

Invertase finds another use in the production of confectionery with liquid or soft centres. These centres are formulated using crystalline sucrose and tiny (about 100 U kg^{-1}, 0.3 p.p.m. (w/w)) amounts of invertase. At this level of enzyme, inversion of sucrose is very slow so the centre remains solid long enough for enrobing with chocolate to be completed. Then, over a period of days or weeks, sucrose hydrolysis occurs and the increase in solubility causes the centres to become soft or liquid, depending on the water content of the centre preparation.

Other enzymes are used as aids to sugar production and refining by removing materials which inhibit crystallisation or cause high viscosity. In some parts of the world, sugar cane contains significant amounts of starch, which becomes viscous, thus slowing filtration processes and making the solution hazy when the sucrose is dissolved. This problem can be overcome by using the most thermostable α-amylases (e.g. Termamyl at about 5 U kg^{-1}) which are entirely compatible with the high temperatures and pH values that prevail during the initial vacuum evaporation stage of sugar production.

Other problems involving dextran and raffinose required the development of new industrial enzymes. A dextran is produced by the action of dextransucrase (EC 2.4.1.5) from *Leuconostoc mesenteroides* on sucrose and found as a slime on damaged cane and beet tissue, especially when processing has been delayed in hot and humid climates. Raffinose, which consists of sucrose with α-galactose attached through its C-4 atom to the 1 position on the fructose residue, is produced at low temperatures in sugar beet. Both dextran and raffinose have the sucrose molecule as part of their structure and both inhibit sucrose crystal growth. This produces plate-like or needle-like crystals which are not readily harvested by equipment designed for the approximately cubic crystals otherwise obtained. Dextran can produce extreme viscosity in process streams and even bring plant to a stop. Extreme dextran problems are frequently solved by the use of fungal dextranases produced from *Penicillium* species. These are used (e.g. 10 U kg^{-1} raw juice, 55 °C, pH 5.5, 1 h) only in times of crisis as they are not sufficiently resistant to thermal denaturation for long-term use and are inactive at high sucrose concentrations. Because only small quantities are produced for use, this enzyme is relatively expensive. An enzyme sufficiently stable for prophylactic use would be required in order to benefit from economies of scale. Raffinose may be hydrolysed to galactose and sucrose by a fungal raffinase (see Chapter 5).

Glucose from cellulose

There is very much more cellulose available, as a potential source of glucose, than starch, yet cellulose is not a significant source of pure glucose. The reasons for this are many, some technical, some commercial. The fundamental reason is that starch is produced in relatively pure forms by plants for use as an easily biodegradable energy and carbon store. Cellulose is structural and is purposefully combined and associated with lignin and pentosans, so as to resist biodegradation; dead trees take several years to decay even in tropical rainforests. A typical waste cellulolytic material contains less than half cellulose, most of the remainder consisting of roughly equal quantities of lignin and pentosans. A combination of enzymes is needed to degrade this mixture. These enzymes are comparatively unstable, of low activity against native lignocellulose and subject to both substrate and product inhibition. Consequently, although many cellulolytic enzymes exist and it is possible to convert pure cellulose to glucose using only enzymes, the cost of this conversion is excessive. The enzymes might be improved by strain selection from the wild or by mutation but problems caused by the physical nature of cellulose are not so amenable to solution. Granular starch is readily stirred in slurries containing 40% (w/v) solids and easily solubilised but, even when pure, fibrous cellulose forms immovable cakes at 10% solids and remains insoluble in all but the most exotic (and enzyme denaturing) solvents. Impure cellulose often contains almost an equal mass of lignin, which is of little or no value as a by-product and is difficult and expensive to remove.

Commercial cellulase preparations from *Trichoderma reesei* consist of mixtures of the synergistic enzymes: (*a*) cellulase (EC 3.2.1.4), an endo-1,4-β-glucanase; (*b*) glucan 1,4-β-glucosidase (EC 3.2.1.74), and exo-1,4-β-glucosidase; and (*c*) cellulose 1,4-β-cellobiosidase (EC 3.2.1.91), an

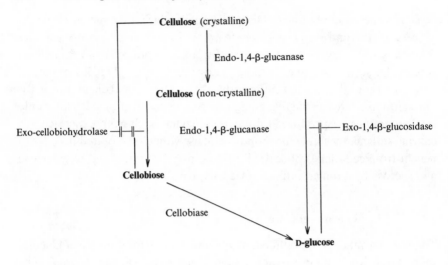

Figure 4.4. Outline of the relationship between the enzyme activities in the hydrolysis of cellulose. ‖ represents inhibitory effects. Endo-1,4-β-glucanase is the rate-controlling activity and may consist of a mixture of enzymes acting on cellulose of different degrees of crystallinity. It acts synergistically with both exo-1,4-β-glucosidase and exo-cellobio-hydrolase. Exo-1,4-β-glucosidase is a product-inhibited enzyme. Exo-cellobiohydrolase is product inhibited and additionally appears to be inactivated on binding to the surface of crystalline cellulose.

exo-cellobiohydrolase (see Figure 4.4). They are used for the removal of relatively small concentrations of cellulose complexes which have been found to interfere in the processing of plant material in, for example, the brewing and fruit juice industries.

Proper economic analysis reveals that cheap sources of cellulose prove to be generally more expensive as sources of glucose than apparently more expensive starch. Relatively pure cellulose is valuable in its own right, as a paper pulp and chipboard raw material, which currently commands a price of over twice that of corn starch. With the increasing world shortage of pulp it cannot be seen realistically as an alternative source of glucose in the foreseeable future. Knowledge of enzyme systems capable of degrading lignocellulose is advancing rapidly but it is unlikely that lignocellulose will replace starch as a source of glucose syrups for food use. It is, however, quite possible that it may be used, in a process involving the simultaneous use of both enzymes and fermentative yeasts, to produce ethanol; the utilisation of the glucose by the yeast removing its inhibitory effect on the enzymes. It should be noted that cellobiose is a non-fermentable sugar and must be hydrolysed by additional β-glucosidase (EC 3.2.1.21, also called cellobiase) for maximum process efficiency (Figure 4.4).

The use of lactases in the dairy industry

Lactose is present at concentrations of about 4.7% (w/v) in milk and the whey (supernatant) left after the coagulation stage of cheese-making. Its presence in milk makes it unsuitable for the majority of the world's adult population, particularly in those areas which have traditionally not had a dairy industry. Real lactose tolerance is confined mainly to peoples whose origins lie in Northern Europe or the Indian subcontinent and is due to 'lactase persistence'; the young of all mammals clearly are able to digest milk but in most cases this ability reduces after weaning. Of the Thai, Chinese and Black American populations, 97%, 90% and 73% respectively, are reported to be lactose intolerant, whereas 84% and 96% of the US White and Swedish populations, respectively, are tolerant. Additionally, and only very rarely, some individuals suffer from inborn metabolic lactose intolerance or lactase deficiency, both of which may be noticed at birth. The need for low-lactose milk is particularly important in food-aid programmes as severe tissue dehydration, diarrhoea and even death may result from feeding lactose-containing milk to lactose-intolerant children and adults suffering from protein-calorie malnutrition. In all these cases, hydrolysis of the lactose to glucose and galactose would prevent the (severe) digestive problems.

Another problem presented by lactose is its low solubility resulting in crystal formation at concentrations above 11% (w/v) (4 °C). This prevents the use of concentrated whey syrups in many food processes as they have an unpleasant sandy texture and are readily prone to microbiological spoilage. Adding to this problem, the disposal of such waste whey is expensive (often punitively so) due to its high biological oxygen demand. These problems may be overcome by hydrolysis of the lactose in whey; the product being about four times as sweet (see Table 4.3), much more soluble and capable of forming concentrated, microbiologically secure, syrups (70% (w/v)).

Lactose may be hydrolysed by lactase, a β-galactosidase.

$$\text{lactose} \xrightarrow{} \text{D-glucose} + \beta\text{-D-galactose} \qquad [4.3]$$

Commercially, it may be prepared from the dairy yeast *Kluyveromyces fragilis* (*K. marxianus* var. *marxianus*), with a pH optimum (pH 6.5–7.0) suitable for the treatment of milk, or from the fungi *Aspergillus oryzae* or *A. niger*, with pH optima (pH 4.5–6.0 and 3.0–4.0, respectively) more suited to whey hydrolysis. These enzymes are subject to varying degrees of product inhibition by galactose. In addition, at high lactose and galactose concentrations, lactase shows significant transferase ability and produces β-1,6-linked galactosyl oligosaccharides.

Lactases are now used in the production of ice cream and sweetened flavoured and condensed milks. When added to milk or liquid whey (2000 U kg^{-1}) and left for about a day at 5 °C about 50% of the lactose is hydrolysed, giving a sweeter product which will not crystallise if condensed or frozen. This method enables otherwise-wasted whey to replace some or all of the skim milk powder used in traditional ice cream recipes. It also improves the 'scoopability' and creaminess of the product. Smaller amounts of lactase may be added to long-life (UHT)-sterilised milk to produce a relatively inexpensive lactose-reduced product (e.g. 20 U kg^{-1}, 20 °C, 1 month of storage). Generally, however, lactase usage has not reached its full potential, as present enzymes are relatively expensive and can only be used at low temperatures.

Enzymes in the fruit juice, wine, brewing and distilling industries

One of the major problems in the preparation of fruit juices and wine is cloudiness due primarily to the presence of pectins. These consist primarily of α-1,4-anhydrogalacturonic acid polymers, with varying degrees of methyl esterification. They are associated with other plant polymers and, after homogenisation, with the cell debris. The cloudiness that they cause is difficult to remove except by enzymic hydrolysis. Such treatment also has the additional benefits of reducing the solution viscosity, increasing the volume of juice produced (e.g. the yield of juice from white grapes can be raised by 15%), subtle but generally beneficial changes in the flavour and, in the case of wine-making, shorter fermentation times. Insoluble plant material is easily removed by filtration, or settling and decantation, once the stabilising effect of the pectins on the colloidal haze has been removed.

Commercial pectolytic enzyme preparations are produced from *Aspergillus niger* and consist of a synergistic mixture of enzymes: (*a*) polygalacturonase (EC 3.2.1.15), responsible for the random hydrolysis of 1,4-α-D-galactosiduronic linkages; (*b*) pectinesterase (EC 3.2.1.11), which releases methanol from the pectyl methyl esters, a necessary stage before the polygalacturonase can act fully (the increase in the methanol content of such treated juice is generally less than the natural concentrations and poses no health risk); (*c*) pectin lyase (EC 4.2.2.10), which cleaves the pectin, by an elimination reaction releasing oligosaccharides with non-reducing terminal 4-deoxy-6-methyl-α-D-galact-4-enuronosyl residues, without the necessity of pectin methyl esterase action; and (*d*) hemicellulase (a mixture of hydrolytic enzymes including: xylan endo-1,3-β-xylosidase, EC 3.2.1.32; xylan 1,4-β-xylosidase, EC 3.2.1.37; and α-L-arabinofuranosidase, EC 3.2.1.55), strictly not a pectinase but its adventitious presence is encouraged in order to reduce hemicellulose levels. The optimal activity of these enzymes is at a pH between 4 and 5 and generally below 50 °C. They are suitable for direct addition to the fruit pulps at levels around 20 U l^{-1} (net activity). Enzymes with improved characteristics of greater heat stability and lower pH optimum are currently being sought.

In brewing, barley malt supplies the major proportion of the enzymes needed for saccharification prior to fermentation. Often other starch-containing material (*adjuncts*) are used to increase the fermentable sugars and reduce the relative costs of the fermentation. Although malt enzymes may also be used to hydrolyse these adjuncts, for maximum economic return extra enzymes are added to achieve their rapid saccharification. It is not necessary nor desirable to saccharify the starch totally, as non-fermentable dextrins are needed to give the drink 'body' and stabilise its foam 'head'. For this reason the saccharification process is stopped, by boiling the 'wort', after about 75% of the starch has been converted into fermentable sugar.

The enzymes used in brewing are needed for saccharification of starch (bacterial and fungal α-amylases), breakdown of barley β-1,4- and β-1,3-linked glucan (β-glucanase) and hydrolysis of protein (neutral protease) to increase the (later) fermentation rate, particularly in the production of high-gravity beer, where extra protein is added. Cellulases are also occasionally used, particularly where wheat is used as adjunct but also to help breakdown the barley β-glucans. Due to the extreme heat stability of the *B. amyloliquefaciens* α-amylase, where this is used the wort must be boiled for a much longer period (e.g. 30 min) to inactivate it prior to fermentation. Papain is used in the later post-fermentation stages of beer-making to prevent the occurrence of protein- and tannin-containing 'chill-haze' otherwise formed on cooling the beer.

Recently, 'light' beers, of lower calorific content, have become more popular. These require a higher degree of saccharification at lower starch concentrations to reduce the alcohol and total solids contents of the beer. This may be achieved by the use of glucoamylase and/or fungal α-amylase during the fermentation.

A great variety of carbohydrate sources are used world wide to produce distilled alcoholic drinks. Many of these contain sufficient quantities of fermentable sugar (e.g. rum from molasses and brandy from grapes), others contain mainly starch and must be saccharified before use (e.g. whiskey from barley malt, corn or rye). In the distilling industry, saccharification continues throughout the fermentation period. In some cases (e.g. Scotch malt whisky manufacture uses barley malt exclusively) the enzymes are naturally present but in others (e.g. grain spirits production) the more heat-stable bacterial α-amylases may be used in the saccharification.

Glucose oxidase and catalase in the food industry

Glucose oxidase is a highly specific enzyme (for D-glucose, but see Chapter 8), from the fungi *Aspergillus niger* and *Penicillium*, which catalyses the oxidation of β-glucose to glucono-1,5-lactone (which spontaneously hydrolyses non-enzymically to gluconic acid) using molecular oxygen and releasing hydrogen peroxide (see reaction scheme [1.1] p. 3). It finds uses in the removal of either glucose or oxygen from foodstuffs in order to improve their storage capability. Hydrogen peroxide is an effective bacteriocide and may be removed, after use, by treatment with catalase (derived from the same fungal fermentations as the glucose oxidase) which converts it to water and molecular oxygen:

$$2H_2O_2 \xrightarrow{\text{catalase}} 2H_2O + O_2 \qquad [4.4]$$

For most large-scale applications the two enzymic activities are not separated. Glucose oxidase and catalase may be used together when net hydrogen peroxide production is to be avoided.

A major application of the glucose oxidase/catalase system is in the removal of glucose from egg-white before drying for use in the baking industry. A mixture of the enzymes is used (165 U kg^{-1}) together with additional hydrogen peroxide (about 0.1% (w/w)) to ensure that sufficient molecular oxygen is available, by catalase action, to oxidise the glucose. Other uses are in the removal of oxygen from the head-space above bottled and canned drinks and reducing non-enzymic browning in wines and mayonnaises.

Medical applications of enzymes

Development of medical applications for enzymes have been at least as extensive as those for industrial applications, reflecting the magnitude of the

Table 4.4 *Some important therapeutic enzymes*

Enzyme	EC number	Reaction	Use
Asparaginase	3.5.1.1	L-Asparagine + $H_2O \longrightarrow$ L-aspartate + NH_3	Leukaemia
Collagenase	3.4.24.3	Collagen hydrolysis	Skin ulcers
Glutaminase	3.5.1.2	L-Glutamine + $H_2O \longrightarrow$ L-glutamate + NH_3	Leukaemia
Hyaluronidase[a]	3.2.1.35	Hyaluronate hydrolysis	Heart attack
Lysozyme	3.2.1.17	Bacterial cell wall hydrolysis	Antibiotic
Rhodanase[b]	2.8.1.1	$S_2O_3^{2-} + CN^- \longrightarrow SO_3^{2-} + SCN^-$	Cyanide poisoning
Ribonuclease	3.1.26.4	RNA hydrolysis	Antiviral
β-Lactamase	3.5.2.6	Penicillin $\longrightarrow$ penicilloate	Penicillin allergy
Streptokinase[c]	3.4.22.10	Plasminogen $\longrightarrow$ plasmin	Blood clots
Trypsin	3.4.21.4	Protein hydrolysis	Inflammation
Uricase[d]	1.7.3.3	Urate + $O_2 \longrightarrow$ allantoin	Gout
Urokinase[e]	3.4.21.31	Plasminogen $\longrightarrow$ plasmin	Blood clots

[a] Hyaluronoglucosaminidase; [b] thiosulphate sulfurtransferase; [c] streptococcal cysteine proteinase; [d] urate oxidase; [e] plasminogen activator.

potential rewards: for example, pancreatic enzymes have been in use since the nineteenth century for the treatment of digestive disorders. The variety of enzymes and their potential therapeutic applications are considerable. A selection of those enzymes which have realised this potential to become important therapeutic agents is shown in Table 4.4. At present, the most successful applications are extracellular: purely topical uses, the removal of toxic substances and the treatment of life-threatening disorders within the blood circulation.

As enzymes are specific biological catalysts, they should make the most desirable therapeutic agents for the treatment of metabolic diseases. Unfortunately a number of factors severely reduces this potential utility:

(*a*) They are too large to be distributed simply within the body's cells. This is the major reason why enzymes have not yet been successfully applied to the large number of human genetic diseases. A number of methods are being developed in order to overcome this by targeting enzymes; as examples, enzymes with covalently attached external β-galactose residues are targeted at hepatocytes and enzymes covalently coupled to target-specific monoclonal antibodies are being used to avoid non-specific side-reactions.

(b) Being generally foreign proteins to the body, they are antigenic and can elicit an immune response which may cause severe and life-threatening allergic reactions, particularly on continued use. It has proved possible to circumvent this problem, in some cases, by disguising the enzyme as an apparently non-proteinaceous molecule by covalent modification. Asparaginase, modified by covalent attachment of polyethylene glycol, has been shown to retain its anti-tumour effect whilst possessing no immunogenicity. Clearly the presence of toxins, pyrogens and other harmful materials within a therapeutic enzyme preparation is totally forbidden. Effectively, this encourages the use of animal enzymes, in spite of their high cost, relative to those of microbial origin.

(c) Their effective lifetime within the circulation may be only a matter of minutes. This has proved easier than the immunological problem to combat, by disguise using covalent modification. Other methods have also been shown to be successful, particularly those involving entrapment of the enzyme within artificial liposomes, synthetic microspheres and red blood cell ghosts. However, although these methods are efficacious at extending the circulatory lifetime of the enzymes, they often cause increased immunological response and additionally may cause blood clots.

In contrast to the industrial use of enzymes, therapeutically useful enzymes are required in relatively tiny amounts but at a very high degree of purity and (generally) specificity. The favoured kinetic properties of these enzymes are low K_m and high V_{max} in order to be maximally efficient even at very low enzyme and substrate concentrations. Thus the sources of such enzymes are chosen with care to avoid any possibility of unwanted contamination by incompatible material and to enable ready purification. Therapeutic enzyme preparations are generally offered for sale as lyophilised pure preparations with only biocompatible buffering salts and mannitol diluent added. The costs of such enzymes may be quite high but still comparable to those of competing therapeutic agents or treatments. As an example, urokinase (a serine protease, see Table 4.4) is prepared from human urine (some genetically engineered preparations are being developed) and used to dissolve blood clots. The cost of the enzyme is about £100 mg^{-1}, with the cost of treatment in a case of lung embolism being about £10000 for the enzyme alone. In spite of this, the market for the enzyme is worth about £70M year^{-1}.

A major potential therapeutic application of enzymes is in the treatment of cancer. Asparaginase has proved to be particularly promising for the treatment of acute lymphocytic leukaemia. Its action depends upon the fact that tumour cells are deficient in aspartate–ammonia ligase activity, which restricts their ability to synthesise the normally non-essential amino acid L-asparagine. Therefore, they are forced to extract it from body fluids. The action of the asparaginase does not affect the functioning of normal cells, which are able to synthesise enough for their own requirements, but reduces the free exogenous concentration and so induces a state of fatal starvation in the susceptible tumour cells. A 60% incidence of complete remission has been reported in a study of almost 6000 cases of acute lymphocytic leukaemia. The enzyme is administered intravenously. It is only effective in reducing asparagine levels within the bloodstream, showing a half-life of about a day (in a dog). This half-life may be increased 20-fold by use of polyethylene glycol-modified asparaginase.

Summary

(*a*) Many important enzyme processes involve the use of freely dissolved enzymes in solution.

(*b*) Proteases are particularly important for their use in food processing, the leather industry and detergents.

(*c*) Starch hydrolysis is the major industrial enzymic bioconversion. Different products and process conditions result in the production of different materials with various properties and uses.

(*d*) A number of enzymes have useful therapeutic properties. Ways are being found to present them successfully to patients.

Bibliography

Adler-Nissen, J. (1985). *Enzymic hydrolysis of food proteins*. London: Elsevier Applied Science.

Bisaria, V. S. & Ghose, T. K. (1981). Biodegradation of cellulosic materials: substrates, microorganisms, enzymes and products. *Enzyme and Microbial Technology*, **3**, 90–104.

Coultate, T. P. (1988). Food: the chemistry of its components, 2nd edn. London: The Royal Society of Chemistry.

Fullbrook, P. D. (1984). The enzymic production of glucose syrups. In *Glucose syrups: science and technology*, ed. S. Z. Dziedzic & M. W. Kearsley, pp. 65–115. London: Elsevier Applied Science.

Gekas, V. & López-Leiva, M. (1985). Hydrolysis of lactose: a literature review. *Process Biochemistry*, **20**, 2–12.

Gusakov, A. V., Sinitsyn, A. P. & Klyosov, A. A. (1985). Kinetics of the enzymic hydrolysis of cellulose. 1. A mathematical model for a batch reactor process. *Enzyme and Microbial Technology*, **70**, 346–52.

Kennedy, J. F., Cabalda, V. M. & White, C. A. (1988). Enzymic starch utilization and genetic engineering. *Trends in Biotechnology*, **6**, 184–9.

Klyosov, A. A. (1986). Enzymic conversion of cellulosic materials to sugar and alcohol: the technology and its implications. *Applied Biochemistry and Biotechnology*, **12**, 249–300.

Novo technical leaflets:
>   *Decolorization of slaughterhouse blood by application of Alcalase 0.6L* (1981).
>   *Termamyl* (1982).
>   *Alcalase* (1984).
>   *Use of Termamyl for starch liquefaction* (1984).
>   *Use of amyloglucosidase Novo and Promozyme*™ *in the production of high dextrose syrup* (1985).

Novo Alle, DK-2880 Bagsvaerd, Denmark: Novo Industri A/S, Enzymes Division.

Peppler, H. J. & Reed, G. (1987). Enzymes in food and feed processing. In *Biotechnology*, vol. 7a *Enzyme technology*, ed. J. F. Kennedy, pp. 547–603. Weinheim: VCH Verlagsgesellschaft mbH.

Reilly, P. J. (1984). Enzymic degradation of starch. In *Starch conversion technology*, ed. G. M. A. Van Beynum & J. A. Roels, pp. 101–42. New York: Marcel Dekker Inc.

Starace, C. & Barfoed, H. C. (1980). Enzyme detergents. *Kirk-Othmer Encyclopedia of Chemical Technology*, 3rd edn, **9**, 138–48.

Towalski, D. (1987). A case study in enzymes: washing powder enzymes. *International Industrial Biotechnology*, Article 77:7:12/1, pp. 198–203.

5   Immobilised enzymes and their use

Enzyme reactors

An enzyme reactor consists of a vessel, or series of vessels, used to perform a desired conversion by enzymic means. A number of important types of such reactor are shown diagrammatically in Figure 5.1. There are several important factors which determine the choice of reactor for a particular process. In general, the choice depends on the cost of a predetermined productivity within the product's specifications. This must be inclusive of the costs associated with substrate(s), downstream processing, labour, depreciation, overheads and process development, in addition to the more obvious costs concerned with building and running the enzyme reactor. Other contributing factors are the form of the enzyme of choice (i.e. free or immobilised), the kinetics of the reaction and the chemical and physical properties of any immobilisation support including whether it is particulate, membranous or fibrous, and its density, compressibility, robustness, particle size and regenerability. Attention must also be paid to the scale of operation, the possible need for pH and temperature control, the supply and removal of gases and the stability of the enzyme, substrate and product. These factors will be discussed in more detail with respect to the different types of reactor.

Batch reactors generally consist of a tank containing a stirrer (*stirred tank reactor*, STR). The tank is normally fitted with fixed baffles which improve the stirring efficiency. A batch reactor is one in which all of the product is removed, as rapidly as is practically possible, after a fixed time. Generally this means that the enzyme and substrate molecules have identical residence times within the reactor, although in some circumstances there may be a need for further additions of enzyme and/or substrate (i.e. fed-batch operation). The operating costs of batch reactors are higher than for continuous processes due to the necessity for the reactors to be emptied and refilled both regularly and often. This procedure is not only expensive in itself but means that there are considerable periods when such reactors are not productive; it also makes uneven demands on both labour and services. STRs can be used for processes involving non-immobilised enzymes, if the consequences of these contami-

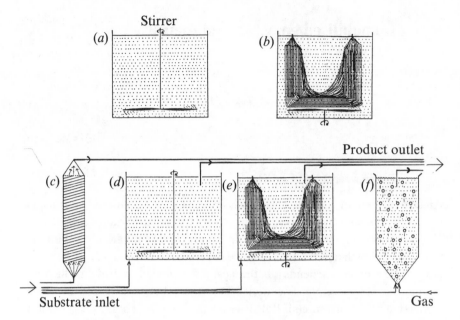

Figure 5.1. Diagrams of various important enzyme reactor types. (*a*) Stirred tank batch reactor (STR), which contains all of the enzyme and substrate(s) until the conversion is complete; (*b*) batch membrane reactor (MR), where the enzyme is held within membrane tubes which allow the substrate to diffuse in and the product to diffuse out. This reactor may often be used in a semicontinuous manner, using the same enzyme solution for several batches; (*c*) packed bed reactor (PBR), also called plug-flow reactor (PFR), containing a settled bed of immobilised enzyme particles; (*d*) continuous flow stirred tank reactor (CSTR) which is a continuously operated version of (*a*); (*e*) continuous flow membrane reactor (CMR) which is a continuously operated version of (*b*); (*f*) fluidised bed reactor (FBR), where the flow of gas and/or substrate keeps the immobilised enzyme particles in a fluidised state. All reactors would additionally have heating/cooling coils (interior in reactors (*a*), and (*d*), and exterior, generally, in reactors (*b*),(*c*), (*e*) and (*f*)) and the stirred reactors may contain baffles in order to increase (reactors (*a*), (*b*), (*d*) and (*e*) or decrease (reactor (*f*)) the stirring efficiency. The continuous reactors ((*c*)–(*f*)) may all be used in a recycle mode where some, or most, of the product stream is mixed with the incoming substrate stream. All reactors may use immobilised enzymes. In addition, reactors (*a*),(*b*) and (*e*) (plus reactors (*d*) and (*f*), if semipermeable membranes are used on their outlets) may be used with the soluble enzyme.

nating the product are not severe. Batch reactors also suffer from pronounced batch-to-batch variations, as the reaction conditions change with time, and may be difficult to scale-up, due to the changing power requirements for efficient fixing. They do, however, have a number of advantageous features. Primary amongst these is their simplicity both in use and in process development. For this reason they are preferred for small-scale production of highly priced products, especially where the same equipment is to be used for a number of different conversions. They offer a closely controllable environment which is useful for slow reactions, where the composition may be accurately monitored, and conditions (e.g. temperature, pH, coenzyme concentrations) varied throughout the reaction. They are also of use when continuous operation of a process proves to be difficult due to the viscous or intractable nature of the reaction mix.

The expected productivity of a batch reactor may be calculated by assuming the validity of the non-reversible Michaelis–Menten reaction scheme with no diffusional control, inhibition or denaturation (see reaction scheme [1.7] (p. 8) and equation (1.7) (p. 9). The rate of reaction (v) may be expressed in terms of the volume of substrate solution within the reactor (Vol_s) and the time (t):

$$v = - Vol_s \frac{d[S]}{dt} = \frac{V_{max}[S]}{K_m + [S]} \tag{5.1}$$

Therefore:

$$\int_0^t \frac{V_{max}}{Vol_s} dt = - \int_{[S]_0}^{[S]_t} \left(1 + \frac{K_m}{[S]}\right) d[S] \tag{5.2}$$

On integrating using the boundary condition that $[S] = [S]_0$ at time (t) = 0:

$$\frac{V_{max}}{Vol_s} t = ([S]_0 - [S]) - K_m \ln\left(\frac{[S]}{[S]_0}\right) \tag{5.3}$$

Let the fractional conversion be X, where:

$$X = \frac{[S]_0 - [S]}{[S]_0} \tag{5.4}$$

therefore:

$$\frac{V_{max}}{Vol_s} t = [S]_0 X - K_m \ln(1 - X) \tag{5.5}$$

The change in fractional conversion and concentrations of substrate and product with time in a batch reactor is shown in Figure 5.2(a).

Membrane reactors

The main requirement for a *membrane reactor* (MR) is a semipermeable membrane which allows the free passage of the product molecules but contains the enzyme molecules. A cheap example of such a membrane is the

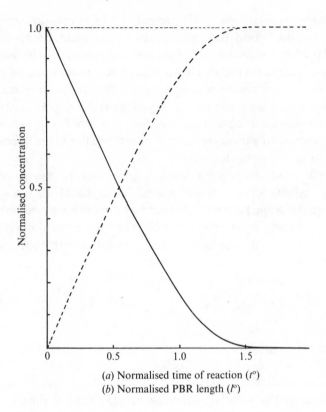

(*a*) Normalised time of reaction (t^o)
(*b*) Normalised PBR length (l^o)

Figure 5.2. This figure shows two related behaviours. (*a*) The change in substrate and product concentrations with time, in a batch reactor. The reaction $S \rightarrow P$ is assumed, with the initial condition $[S]_0/K_m = 10$. The concentrations of substrate (————) and product (– – – –) are both normalised with respect to $[S]_0$. The normalised time (i.e. $t^0 = tV_{max}/[S]_0$) is relative to the time ($t^0 = 1$) that would be required to convert all the substrate if the enzyme acted at V_{max} throughout, the actual time for complete conversion being longer due to the reduction in the substrate concentration at the reaction progresses. The dashed line also indicates the variation of the fractional conversion (X) with t^0.

(*b*) The change in substrate and product concentrations with reactor length for a PBR. The reaction $S \rightarrow P$ is assumed with the initial condition, $[S]_0/K_m = 10$. The concentrations of substrate (————) and product (– – – –) are both normalised with respect to $[S]_0$. The normalised reactor length (i.e. $l^0 = lv_{max}/F$, where V_{max} is the maximum velocity for unit reactor length and l is the reactor length) is relative to the length (i.e. when $l^0 = 1$) that contains sufficient enzyme to convert all the substrate at the given flow rate if the enzyme acted at its maximum velocity throughout; the actual reactor length necessary for complete conversion being longer due to the reduction in the substrate concentration as the reaction progresses. l^0 may be considered as the relative position within a PBR or the reactor's absolute length.

dialysis membrane used for removing low molecular weight species from protein preparations. The usual choice for a membrane reactor is a hollow-fibre reactor consisting of a preformed module containing hundreds of thin tubular fibres each having a diameter of about 200 μm and a membrane thickness of about 50 μm. Membrane reactors may be used in either batch or continuous mode and allow the easy separation of the enzyme from the product. They are normally used with soluble enzymes, avoiding the costs and problems associated with other methods of immobilisation and some of the diffusion limitations of immobilised enzymes. If the substrate is able to diffuse through the membrane, it may be introduced to either side of the membrane with respect to the enzyme, otherwise it must be within the same compartment as the enzyme, a configuration that imposes a severe restriction on the flow rate through the reactor, if used in continuous mode. Due to the ease with which membrane reactor systems may be established, they are often used for production on a small scale (g to kg), especially where a multi-enzyme pathway or coenzyme regeneration is needed. They allow the easy replacement of the enzyme in processes involving particularly labile enzymes and can also be used for biphasic reactions (see Chapter 7). The major disadvantage of these reactors concerns the cost of the membranes and their need to be replaced at regular intervals.

The kinetics of membrane reactors are similar to those of the batch STR, in batch mode, or the CSTR, in continuous mode (see below). Deviations from these models occur primarily in configurations where the substrate stream is on the side of the membrane opposite to the enzyme and the reaction is severely limited by its diffusion through the membrane and the products' diffusion in the reverse direction. Under these circumstances the reaction may be even more severely affected by product inhibition or the limitations of reversibility than is indicated by these models.

Continuous flow reactors

The advantages of immobilised enzymes as processing catalysts are most markedly appreciated in continuous flow reactors. In these, the average residence time of the substrate molecules within the reactor is far shorter than that of the immobilised-enzyme catalyst. This results in a far greater productivity from a fixed amount of enzyme than is achieved in batch processes. It also allows the reactor to handle substrates of low solubility by permitting the use of large volumes containing low concentrations of substrate. The constant reaction conditions may be expected to result in a purer and more reproducible product. There are two extremes of process kinetics in relation to continuous flow reactors; (*a*) the ideal *continuous flow stirred tank reactor* (CSTR), in which the reacting stream is completely and rapidly mixed with the whole of the reactor contents and the enzyme contacts low substrate and high product concentrations; and (*b*) the ideal continuously operated *packed bed reactor* (PBR), where no mixing takes place and the enzyme contacts high substrate and low product concentrations. The properties of the continuously operated *fluidised bed reactor* (FBR) lie, generally, somewhere between these extremes. An ordered series of CSTRs or FBRs may approximate, in use where the outlet of one reactor forms the inlet to the next reactor, to an equivalent PBR.

The transport of momentum, heat and mass in these continuous reactors are important factors contributing to the resultant productivity. They are due to fluid flow and molecular and turbulent motions, and often described by means of a number of empirical relationships, involving dimensionless numbers. The most important of these is the *Reynolds number* (*Re*), which relates the inertial force due to the flow of solution to the viscous force resisting that flow. Low *Re* indicates streamlined flow whereas higher *Re* indicates progressively more turbulence, there being a critical value for *Re*, dependent on the configuration of the system, at which there is a transition from streamlined flow to turbulent flow. *Re* is defined in terms of Lf_m/η or Lf/v, where L is the characteristic length of the system, f_m is the mass flow rate (g m^{-2} s^{-1}), f is the fluid velocity (m s^{-1}), η is the dynamic viscosity (g m^{-1} s^{-1}) and v is the kinematic viscosity (m^2 s^{-1}). Examples of the characteristic lengths in these definitions of *Re* are the particle diameter for flow past spherical particles, and the tube diameter for flow within hollow tubes. The value of the (characteristic length of the system) × (fluid velocity) for a stirred tank system may be taken as the (stirrer speed) × (stirrer diameter)^2. *Re* is, therefore, higher at high flow rates and low viscosities. Another dimensionless number which is of use in describing and comparing continuous enzyme reactors is the *Le Goff number* (*Lf*), which expresses the efficiency with which the energy dissipated in producing the fluid flow is used to transport material (and heat) to the catalytic surface. Low values for *Lf* indicate a relatively high energetic and financial cost, in achieving contact between the catalytically active immobilised-enzyme surface and the substrate stream, and the consequent reduction in any external diffusion limitations. The *Lf* is higher for low pressure drops through the reactor, high flow rates and high conversions. The relationship between the *Lf* and the *Re* is shown schematically for a PBR and a CSTR in Figure 5.3, where it may be seen that the PBR is more efficient at fairly low *Re* but far less so at higher *Re*, reflecting the necessity for stirring the CSTR even at low flow rates and the increased backpressure in PBRs at high flow rates.

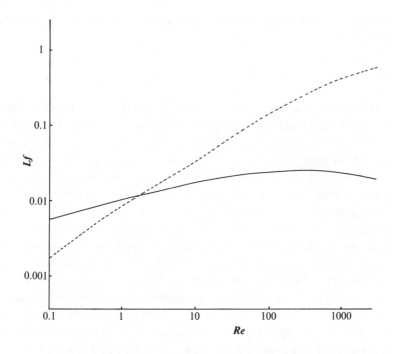

Figure 5.3. A diagram comparing the variation of *Re* with *Lf* for a PBR
(———) and a CSTR (––––––). An FBR behaves similarly to the CSTR.

Packed bed reactors

The most important characteristic of a PBR is that material flows through the
reactor as a plug; they are also called *plug-flow reactors* (PFR). Ideally, all of
the substrate stream flows at the same velocity, parallel to the reactor axis,
with no back-mixing. All material present at any given reactor cross-section
has had an identical residence time. The longitudinal position within the
PBR is, therefore, proportional to the time spent within the reactor; all
product emerging with the same residence time and all substrate molecules
having an equal opportunity for reaction. The conversion efficiency of a
PBR, with respect to its length, behaves in a manner similar to that of a
well-stirred batch reactor with respect to its reaction time (Figure 5.2(b)).
Each volume element behaves as a batch reactor as it passes through the
PBR. Any required degree of reaction may be achieved by use of an ideal
PBR of suitable length.

The flow rate (F) is equivalent to Vol_s/t for a batch reactor. Therefore,
equation (5.5) may be converted to represent an ideal PBR, given the

assumption, not often realised in practice, that there are no diffusion limitations:

$$\frac{V_{max}}{F} = [S]_0 X - K_m \ln(1 - X) \qquad (5.6)$$

In order to produce ideal plug-flow within PBRs, a turbulent flow regime is preferred to laminar flow, as this causes improved mixing and heat transfer normal to the flow and reduced axial back-mixing. Achievement of high enough *Re* may, however, be difficult due to unacceptably high feed rates. Consequent upon the plug-flow characteristic of the PBR is that the substrate concentration is maximised, and the product concentration minimised, relative to the final conversion at every point within the reactor; the effectiveness factor being high on entry to the reactor and low close to the exit. This means that PBRs are the preferred reactors, all other factors being equal, for processes involving product inhibition, substrate activation and reaction reversibility. At low *Re* the flow rate is proportional to the pressure drop across the PBR. This pressure drop is, in turn, generally found to be proportional to the bed height, the linear flow rate and dynamic viscosity of the substrate stream and $(1 - \epsilon)^2/\epsilon^3$ (where ϵ is the porosity of the reactor; i.e. the fraction of the PBR volume taken up by the liquid phase), but inversely proportional to the cross-sectional area of the immobilised enzyme pellets. In general PBRs are used with fairly rigid immobilised-enzyme catalysts (1–3 mm diameter), because excessive increases in this flow rate may distort compressible or physically weak particles. Particle deformation results in reduced catalytic surface area of particles contacting the substrate-containing solution, poor external mass transfer characteristics and a restriction to the flow, causing increased pressure drop. A vicious circle of increased back-pressure, particle deformation and restricted flow may eventually result in no flow at all through the PBR.

PBRs behave as deep-bed filters with respect to the substrate stream. It is necessary to use a guard bed if plugging of the reactor by small particles is more rapid than the biocatalysts' deactivation. They are also easily fouled by colloidal or precipitating material. The design of PBRs does not allow for control of pH, by addition of acids or bases, or for easy temperature control where there is excessive heat output, a problem which may be particularly noticeable in wide reactors (> 15 cm diameter). Deviations from ideal plug-flow are due to back-mixing within the reactors, the resulting product streams having a distribution of residence times. In an extreme case, back-mixing may result in the kinetic behaviour of the reactor approximating to that of the CSTR (see below), and the consequent difficulty in achieving a high degree of conversion. These deviations are caused by channelling, where some substrate passes through the reactor more rapidly, and hold-up, which involves stagnant areas with negligible flow rate. Channels may form in the reactor bed due to excessive pressure drop, irregular packing or uneven application of the substrate stream, causing flow rate differences across the bed. The use of an uniformly sized catalyst in a reactor with an upwardly flowing substrate stream reduces the chance and severity of non-ideal behaviour.

Continuous flow stirred tank reactors

This reactor consists of a well-stirred tank containing the enzyme, which is normally immobilised. The substrate stream is continuously pumped into the reactor at the same time as the product stream is removed. If the reactor is behaving in an ideal manner, there is total back-mixing and the product stream is identical with the liquid phase within the reactor and invariant with respect to time. Some molecules of substrate may be removed rapidly from the reactor, whereas others may remain for substantial periods. The distribution of residence times for molecules in the substrate stream is shown in Figure 5.4.

The CSTR is an easily constructed, versatile and cheap reactor, which allows simple catalyst charging and replacement. Its well-mixed nature permits straightforward control over the temperature and pH of the reaction, and the supply or removal of gases. CSTRs tend to be rather large as they need to be efficiently mixed. Their volumes are usually about five to ten times the volume of the contained immobilised enzyme. This, however, has the advantage that there is very little resistance to the flow of the substrate stream, which may contain colloidal or insoluble substrates, so long as the insoluble particles are not able to sweep the immobilised enzyme from the reactor. The mechanical nature of the stirring limits the supports for the immobilised enzymes to materials which do not easily disintegrate to give 'fines' which may enter the product stream. However, fairly small particles (down to about 10 μm diameter) may be used, if they are sufficiently dense to stay within the reactor. This minimises problems due to diffusional resistance.

An ideal CSTR has complete back-mixing resulting in a minimisation of the substrate concentration, and a maximisation of the product concentration, relative to the final conversion, at every point within the reactor, the effectiveness factor being uniform throughout. Thus, CSTRs are the preferred reactors, everything else being equal, for processes involving substrate inhibition or product activation. They are also useful where the substrate stream contains an enzyme inhibitor, as it is diluted within the

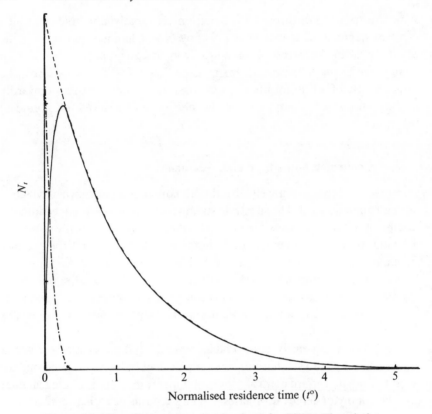

Figure 5.4. The residence time distribution of a CSTR. The relative number of molecules resident within the reactor for a particular time N_t is plotted against the normalised residence time (i.e. tF/V, where V is the reactor volume, and F is the flow rate; it is the time relative to that required for one reactor volume to pass through the reactor). The residence time distribution of non-reacting media molecules (------, which obeys the relationship $[M_t] = [M_o]e^{-t \times F/V}$, where $[M]$ is the concentration of media molecules, giving a half-life for remaining in the reactor of $\ln(2) \times V/F, = 0.7V/F$), product (———) and substrate (–·–·–·–) are shown. The reaction $S \rightarrow P$ is assumed, and substrate molecules which have long residence times are converted into product, the average residence time of the product being greater than that for the substrate. The composition of the product stream is identical with that of the liquid phase within the reactor. This composition may be calculated from the relative areas under the curves and, in this case, represents a 90% conversion. Under continuous operating conditions (operating time $> 4V/F$), the mean residence time within the reactor is V/F. However, it may be noted from the graph that only a few molecules have a residence time close to this value (only 7% between $0.9V/F$ and $1.1V/F$) whereas 20% of the molecules have residence times of less than $0.1V/F$ or greater than $2.3V/F$. It should be noted that 100% of the molecules in an equivalent ideal PBR might be expected to have residence times equal to their mean residence time.

The y-axis is labelled N_t and the x-axis is labelled Normalised residence time (t^o), with markings at 0, 1, 2, 3, 4, 5.

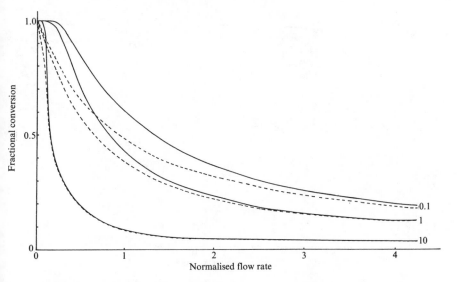

Figure 5.5. Comparison of the changes in fractional conversion with flow rate between the PBR (————) and CSTR (------) at different values of $[S]_0/K_m$ (10, 1 and 0.1, higher $[S]_0/K_m$ giving the higher curves). The flow rate is normalised with respect to the reactor's volumetric enzyme content ($= F \times K_m/V_{max}$). It can be seen that there is little difference between the two reactors at faster flow rates and lower conversions, especially at high values of $[S]_0/K_m$.

reactor. This effect is most noticeable if the inhibitor concentration is greater than the inhibition constant and $[S]_0/K_m$ is low for competitive inhibition or high for uncompetitive inhibition, when the inhibitor dilution has more effect than the substrate dilution. Deviations from ideal CSTR behaviour occur when there is a less effective mixing regime and may generally be overcome by increasing the stirrer speed, decreasing the solution viscosity or biocatalyst concentration or by more effective reactor baffling.

The rate of reaction within a CSTR can be derived from a simple mass balance to be the flow rate (F) times the difference in substrate concentration between the reactor inlet and outlet. Hence:

$$F([S]_0 - [S]) = \frac{V_{max}[S]}{K_m + [S]} \tag{5.7}$$

therefore:

$$\frac{V_{max}}{F} = K_m \frac{([S]_0 - [S])}{[S]} + ([S]_0 - [S]) \tag{5.8}$$

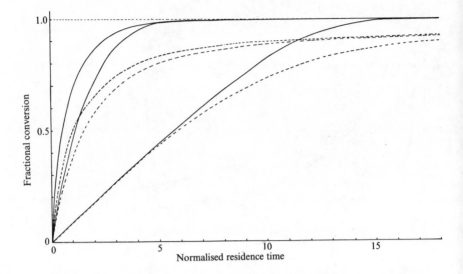

Figure 5.6. Comparison of the changes in fractional conversion with residence time between the PBR (——) and CSTR (-----) at different values of $[S]_0/K_m$ (10, 1 and 0.1; higher $[S]_0/K_m$ values giving the lower curves). The residence time is the reciprocal of the normalised flow rate (see Figure 5.5). If the flow rate is unchanged then the 'normalised residence time' may be thought of as the reactor volume needed to produce the required degree of conversion.

from equation (5.4):

$$\frac{[S]_0 - [S]}{[S]} = \frac{X}{1 - X} \tag{5.9}$$

therefore:

$$\frac{V_{max}}{F} = [S]_0 X + K_m \frac{X}{1 - X} \tag{5.10}$$

This equation should be compared with that for the PBR (equation (5.6)). Together these equations can be used for comparing the productivities of the two reactors (Figures 5.5 and 5.6).

Equations describing the behaviour of CSTRs and PBRs utilising reversible reactions or undergoing product or substrate inhibition can be derived in a similar manner, using equations (1.68), (1.85) and (1.96) rather than (1.8):

Substrate-inhibited PBR:

$$V_{max}/F = X[S]_0 - K_m \ln(1 - X) + [S]_0^2 (2X - X^2)/2K_s \tag{5.11}$$

Substrate-inhibited CSTR:

$$V_{max}/F = X[S]_0 + K_m[X/(1 - X)] + [S]_0^2(X - X^2)/K_s \qquad (5.12)$$

Product-inhibited PBR:

$$V_{max}/F = X[S]_0(1 - K_m/K_P) - K_m\ln(1 - X)(1 + [S]_0/K_P) \qquad (5.13)$$

Product-inhibited CSTR:

$$V_{max}/F = X[S]_0 + K_m\{X/(1 - X)\}(1 + X[S]_0/K_P) \qquad (5.14)$$

Reversible reaction in a PBR:

$$V/F = X[S^*] - K \ln(1 - X) \qquad (5.15)$$

Reversible reaction in a CSTR:

$$V/F = X[S^*] + K\{X/(1 - X)\} \qquad (5.16)$$

X in equations (5.15) and (5.16) is the fractional conversion for a reversible reaction.

$$X = \frac{[S]_0 - [S]}{[S]_0 - [S]_\infty} \qquad (5.17)$$

The meaning of other symbols used in equations (5.11)–(5.17) are given in Chapter 1. These equations may be used to compare the size of PBR and CSTR necessary to achieve the same conversion under various conditions (Figure 5.7). Another useful parameter for comparing these reactors is the productivity. This can be derived for each reactor assuming a first-order inactivation of the enzyme (equation (1.26), p. 21)). Combined with equation (5.6) for PBR, or (5.10) for CSTR, the following relationships are obtained on integration:

$$\text{PBR:} \quad \ln\frac{\{[S]_0X_0 - K_m\ln(1 - X_0)\}}{\{[S]_0X_t - K_m\ln(1 - X_t)\}} = k_d t \qquad (5.18)$$

$$\text{CSTR:} \quad \ln\frac{\{[S]_0X_0 + K_m(X_0/(1 - X_0))\}}{\{[S]_0X_t + K_m(X_t/(1 - X_t))\}} = k_d t \qquad (5.19)$$

where k_d is the first-order inactivation constant (i.e. k_{d1} in equation (1.25)), and the fractional conversion subscripts refer to time = 0 or t. The change in productivity (Figure 5.8) and fractional conversion (Figure 5.9) of these reactors with time can be compared using these equations.

These reactors may be operated for considerably longer periods than that determined by the inactivation of their contained immobilised enzyme, particularly if they are capable of high conversion at low substrate concentrations (Figure 5.9). This is independent of any enzyme stabilisation and is

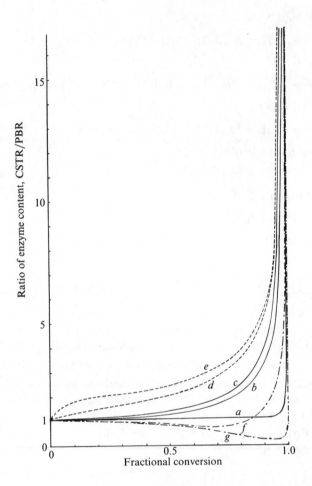

Figure 5.7. Comparison of the ratio, of the enzyme content in a CSTR to that in a PBR, necessary to achieve various degrees of conversion for a range of process conditions. The actual size of the CSTR will be five to ten times greater than indicated due to the necessity of maintaining stirring within the vessel. ——— Uninhibited reaction; – – – – product inhibited; –·–·–· substrate inhibited. (curve *a*) $[S]_0/K_m = 100$; (curve *b*) $[S]_0/K_m = 1$; (curve *c*) $[S]_0/K_m < 0.01$; (curve *d*) $[S]_0/K_m = 1$, product inhibited $K_P/K_m = 0.1$; (curve *e*) $[S]_0/K_m = 1$, product inhibited $K_P/K_m = 0.01$; (curve *f*) $[S]_0/K_m = 1$, substrate inhibited $[S]_0/K_S = 10$; (curve *g*) $[S]_0/K_m = 100$; substrate inhibited $[S]_0/K_S = 10$. The size of a CSTR becomes prohibitively large at high conversions (e.g. using curve *b*, a CSTR contains three times the enzyme in a PBR to achieve a 90% conversion, but this increases to 18 times for 99% conversion. The difference between the two types of reactor is increased if the effectiveness factor (η) is less than one due to diffusional effects.

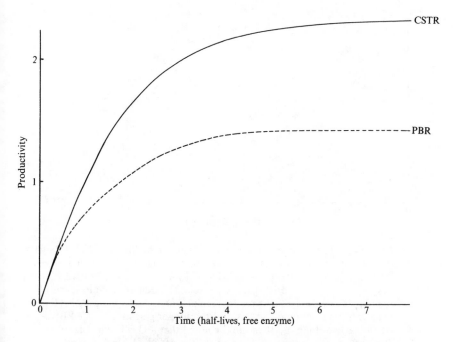

Figure 5.8. The change in productivity of a PBR (-----) and CSTR
(———) with time, assuming an initial fractional conversion (X_0) = 0.99
and $[S]_0/K_m = 100$. The units of time are half-lives of the free enzyme
($t_{1/2}, = \ln(2)/k_d$) and the productivity is given in terms of ($FXt_{1/2}$).
Although the overall productivity is 1.6 times greater for a CSTR than a
PBR, it should be noted that the CSTR contains 1.9 times more enzyme.

simply due to such reactors initially containing large amounts of redundant
enzyme.

In general, there is little or no back-pressure to increased flow rates
through the CSTR. Such reactors may be started up as batch reactors until
the required degree of conversion is reached, when the process may be made
continuous. CSTRs are not generally used in processes involving high
conversions but a train of CSTRs may approach the PBR performance. This
train may be a number (greater than three) of reactors connected in series or a
single vessel divided into compartments, in order to minimise back-mixing.
CSTRs may be used with soluble rather than immobilised enzyme if an
ultrafiltration membrane is used to separate the reactor output stream from
the reactor contents. This causes a number of process difficulties, including
concentration polarization or inactivation of the enzyme on the membrane,
but may be preferable in order to achieve a combined reaction and separation
process or where a suitable immobilised enzyme is not readily available.

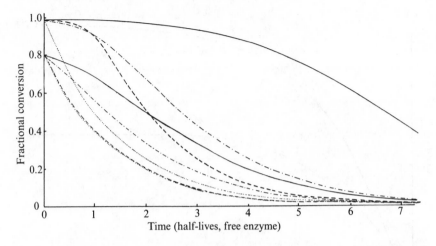

Figure 5.9. The change in fractional conversion of PBRs and CSTRs with time, assuming initial fractional conversion (X_0) of 0.99 or 0.80. ——— CSTR, $[S]_0/K_m = 0.01$; ––––– CSTR, $[S]_0/K_m = 100$; –·–·–· PBR $[S]_0/K_m = 0.01$; ...... PBR $[S]_0/K_m = 100$. The time is given in terms of the half-life of the free enzyme ($t_{1/2} = \ln(2)/k_d$). Although the CSTR maintains its fractional conversion for a longer period than the PBR, particularly at high X_0. It should be noted that a CSTR capable of $X_0 = 0.99$ at a substrate feed concentration of $[S]_0/K_m = 0.01$ contains 22 times more enzyme than an equivalent PBR, but yields only 2.2 times more product. The initial stability in the fractional conversion over a considerable period of time, due to the enzyme redundancy, should not be confused with any effect due to stabilisation of the immobilised enzyme.

Fluidised bed reactors

These reactors generally behave in a manner intermediate between CSTRs and PBRs. They consist of a bed of immobilised enzyme which is fluidised by the rapid upwards flow of the substrate stream alone or in combination with a gas or secondary liquid stream, either of which may be inert or contain material relevant to the reaction. A gas stream is usually preferred as it does not dilute the product stream. There is a minimum fluidisation velocity needed to achieve bed expansion, which depends upon the size, shape, porosity and density of the particles and the density and viscosity of the liquid. This minimum fluidisation velocity is generally fairly low (about 0.2–1.0 cm s^{-1}) as most immobilised-enzyme particles have densities close to that of the bulk liquid. In this case the relative bed expansion is proportional to the superficial gas velocity and inversely proportional to the square root of the reactor diameter. Fluidising the bed requires a large power input but, once fluidised, there is little further energetic input needed to increase the flow rate of the substrate stream through the reactor (Figure 5.3). At high flow rates and low reactor diameters almost ideal plug-flow characteristics may be achieved. However, the kinetic performance of the FBR normally lies between that of the PBR and the CSTR, as the small fluid linear velocities allowed by most biocatalytic particles causes a degree of back-mixing that is often substantial, although never total. The actual design of the FBR will determine whether it behaves in a manner which is closer to that of a PBR or CSTR (see Figures 5.5–5.9). It can, for example, be made to behave in a manner very similar to that of a PBR, if it is baffled in such a way that substantial backmixing is avoided. FBRs are chosen when these intermediate characteristics are required, e.g. where a high conversion is needed but the substrate stream is colloidal or the reaction produces a substantial pH change or heat output. They are particularly useful if the reaction involves the utilisation or release of gaseous material.

The FBR is normally used with fairly small immobilised enzyme particles (20–40 μm diameter) in order to achieve a high catalytic surface area. These particles must be sufficiently dense, relative to the substrate stream, that they are not swept out of the reactor. Less-dense particles must be somewhat larger. For efficient operation the particles should be of nearly uniform size, otherwise a non-uniform biocatalytic concentration gradient will be formed up the reactor. FBRs are usually tapered outwards at the exit to allow for a wide range of flow rates. Very high flow rates are avoided as they cause channelling and catalyst loss. The major disadvantage of development of a FBR process is the difficulty in scaling-up these reactors. PBRs allow scale-up factors of greater than 50 000 but, because of the markedly different fluidisation characteristics of different sized reactors, FBRs can only be scaled-up by a factor of 10–100 each time. In addition, changes in the flow rate of the substrate stream causes complex changes in the flow pattern within these reactors which may have consequent unexpected effects upon the conversion rate.

Immobilised-enzyme processes

Immobilised-enzyme systems are used where they offer cost advantages to users on the basis of total manufacturing costs. The plant size needed for continuous processes is two orders of magnitude smaller than that required for batch processes using free enzymes. The capital costs are, therefore, considerably smaller and the plant may be prefabricated cheaply off-site. Immobilised enzymes offer greatly increased productivity on an enzyme weight basis and also often provide process advantages (see Chapter 3). Currently used immobilised-enzyme processes are given in Table 5.1.

Table 5.1 *Some of the more important industrial uses of immobilised enzymes*

Enzyme	EC number	Product
Aminoacylase	3.5.1.14	L-Amino acids
Aspartate ammonia-lyase	4.3.1.1	L-Aspartic acid
Aspartate 4-decarboxylase	4.1.1.12	L-Alanine
Cyanidase	3.5.5.x	Formic acid (from waste cyanide)
Glucoamylase	3.2.1.3	D-Glucose
Glucose isomerase	5.3.1.5	High-fructose corn syrup
Histidine ammonia-lyase	4.3.1.3	Urocanic acid
Hydantoinase[a]	3.5.2.2	D- and L-amino acids
Invertase	3.2.1.26	Invert sugar
Lactase	3.2.1.23	Lactose-free milk and whey
Lipase	3.1.1.3	Cocoa butter substitutes
Nitrile hydratase	4.2.1.x	Acrylamide
Penicillin amidases	3.5.1.11	Penicillins
Raffinase	3.2.1.22	Raffinose-free solutions
Thermolysin	3.2.24.4	Aspartame

[a] Dihydropyrimidinase.

High-fructose corn syrups (HFCS)

With the development of glucoamylase in the 1940s and 1950s it became a straightforward matter to produce high *DE* glucose syrups. However, these have shortcomings as objects of commerce: D-glucose has only about 70% of the sweetness of sucrose, on a weight basis, and is comparatively insoluble. Batches of 97 *DE* glucose syrup at the final commercial concentration (71% (w/w)) must be kept warm to prevent crystallisation or diluted to concentrations that are microbiologically insecure. Fructose is 30% sweeter than sucrose, on a weight basis, and twice as soluble as glucose at low temperatures so a 50% conversion of glucose to fructose overcomes both problems giving a stable syrup that is as sweet as a sucrose solution of the same concentration (see Table 4.3, p. 155). The isomerisation is possible by chemical means but not economical, giving tiny yields and many by-products (e.g. 0.1 M glucose 'isomerised' with 1.22 M KOH at 5 °C under nitrogen for 3.5 months gives a 5% yield of fructose but only 7% of the glucose remains unchanged, the majority being converted to various hydroxy acids).

One of the triumphs of enzyme technology so far has been the development of 'glucose isomerase'. Glucose is normally isomerised to fructose during glycolysis but both sugars are phosphorylated. The use of this phosphohexose isomerase may be ruled out as a commercial enzyme because of the cost of the ATP needed to activate the glucose and because two other enzymes (hexokinase and fructose-6-phosphatase) would be needed to complete the conversion. Only an isomerase that would use underivatised glucose as its substrate would be commercially useful but, until the late 1950s, the existence of such an enzyme was not suspected. At about this time, enzymes were found that catalyse the conversion of D-xylose to an equilibrium mixture of D-xylulose and D-xylose in bacteria. When supplied with cobalt ions, these xylose isomerases were found to isomerise α-D-glucopyranose to α-D-fructo-furanose (see reaction scheme [1.5], p. 4), equilibration from the more abundant β-D-glucopyranose and to the major product β-D-fructopyranose occurring naturally and non-enzymically. Now it is known that several genera of microbes, mainly bacteria, can produce such glucose isomerases. The commercial enzymes are produced by *Actinoplanes missouriensis*, *Bacillus coagulans* and various *Streptomyces* species; as they have specificities for glucose and fructose which are not much different from that for xylose and ways are being found to avoid the necessity of xylose as inducer, these should perhaps now no longer be considered as xylose isomerases. They are remarkably amenable enzymes in that they are resistant to thermal denaturation and will act at very high substrate concentrations, which have the additional benefit of substantially stabilising the enzymes at higher operational temperatures. The vast majority of glucose isomerases are retained within the cells that produce them but need not be separated and purified before use.

All glucose isomerases are used in immobilised forms. Although different immobilisation methods have been used for enzymes from different organisms, the principles of use are very similar. Immobilisation is generally by cross-linking with glutaraldehyde, plus in some cases a protein diluent, after cell lysis or homogenisation.

Originally, immobilised glucose isomerase was used in a batch process. This proved to be costly as the relative reactivity of fructose during the long residence times gave rise to significant by-product production. Also, difficulties were encountered in the removal of the added Mg^{2+} and Co^{2+} and the recovery of the catalyst. Nowadays most isomerisation is performed in PBRs (Table 5.2). They are used with high substrate concentration (35–45% dry solids, 93–97% glucose) at 55–60 °C. The pH is adjusted to 7.5–8.0 using sodium carbonate and magnesium sulphate is added to maintain enzyme activity (Mg^{2+} and Co^{2+} are cofactors). The Ca^{2+} concentration of the glucose feedstock is usually about 25 μM, left from previous processing, and this presents a problem. Ca^{2+} competes successfully for the Mg^{2+} binding site on the enzyme, causing inhibition. At this level the substrate stream is normally made 3 mM with respect to Mg^{2+}. At higher concentrations of

Table 5.2   *Comparison of glucose isomerisation methods*

Parameter	Batch (soluble GI)	Batch (immobilised GI)	Continuous (PBR)
Reactor volume (m³)	1100	1100	15
Enzyme consumption (tonnes)	180	11	2
Activity, half-life (h)	30	300	1500
Active life, half-lives	0.7	2	3
Residence time (h)	20	20	0.5
Co^{2+} (tonnes)	2	1	0
Mg^{2+} (tonnes)	40	40	7
Temperature (°C)	65	65	60
pH	6.8	6.8	7.6
Colour formation (A_{420})	0.7	0.2	<0.1
Product refining	Filtration	—	—
	C-treatment[a]	C-treatment	C-treatment
	Cation exchange	Cation exchange	—
	Anion exchange	Anion exchange	—
Capital, labour and energy costs, £ tonne⁻¹	5	5	1
Conversion cost, £ tonne⁻¹	500	30	5

All processes start with 45% (w/w) glucose syrup *DE* 97 and produce 10 000 tonnes per month of 42% fructose dry syrup. Some of the improvement that may be seen for PBR productivity is due to the substantial development of this process.
[a] Treatment with activated carbon.

calcium a $Mg^{2+} : Ca^{2+}$ ratio of 12 is recommended. Excess Mg^{2+} is uneconomic as it adds to the purification as well as the isomerisation costs. The need for Co^{2+} has not been eliminated altogether, but the immobilisation methods now used fix the cobalt ions so that none needs to be added to the substrate streams.

It is essential for efficient use of immobilised glucose isomerase that the substrate solution is adequately purified so that it is free of insoluble material and other impurities that might inactivate the enzyme by chemical (inhibitory) or physical (pore-blocking) means. In effect, this means that glucose produced by acid hydrolysis cannot be used, as its low quality necessitates extensive and costly purification. Insoluble material is removed by filtration, sometimes after treatment with flocculants, and soluble materials are removed by ion exchange resins and activated carbon beads. This done, there still remains the possibility of inhibition due to oxidised by-products caused by molecular oxygen. This may be removed by vacuum de-aeration of the substrate at the isomerisation temperature or by the addition of low concentrations (< 50 p.p.m.) of sulphite.

At equilibrium at 60 °C about 51% of the glucose in the reaction mixture is

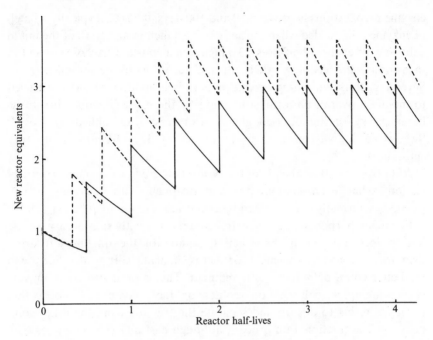

Figure 5.10. Diagram showing the production rate of a seven-column
PBR facility on start-up, assuming exponential decay of reactor activity.
The columns are brought into use one at a time. At any time a maximum
of six PBRs are operating in parallel, whilst the seventh, exhausted,
reactor is being refilled with fresh biocatalyst. ——— PBR activities
allowed to decay through three half-lives (to 12.5% initial activity) before
replacement. The final average productivity is 2.51 times the initial
productivity of one column. – – – – PBR activities allowed to decay
through two half-lives (to 25% initial activity) before replacement. The
final average productivity is 3.23 times the initial productivity of one
column. It may be seen that the final average production rate is higher
when the PBRs are individually operated for shorter periods but this 29%
increase in productivity is achieved at a cost of 50% more enzyme, due to
the more rapid replacement of the biocatalyst in the PBRs. A shorter
PBR operating time also results in a briefer start-up period and a more
uniform productivity.

converted to fructose (see Chapter 1 for a full discussion of such reversible
reactions). However, because of the excessive time taken for equilibrium to
be attained and the presence of oligosaccharides in the substrate stream, most
manufacturers adjust flow rates so as to produce 42–46% (w/w) fructose
(leaving 47–51% (w/w) glucose). To produce 100 tonnes (dry substance) of
42% HFCS per day, an enzyme bed volume of about 4 m^3 is needed. Activity
decreases, following a first-order decay equation. The half-life of most

enzyme preparations is between 50 and 100 days at 55 °C. Typically a batch of enzyme is discarded when the activity has fallen to an eighth of the initial value (i.e. after three half-lives). To maintain a constant fructose content in the product, the feed flow rate is adjusted according to the enzyme activity. Several reactors containing enzyme preparations of different ages are needed to maintain overall uniform production by the plant (Figure 5.10). In its lifetime 1 kg of immobilised glucose isomerase (exemplified by Novo's Sweetzyme T) will produce 10–11 tonnes of 42% fructose syrup (dry substance).

After isomerisation, the pH of the syrup is lowered to 4–5 and it is purified by ion-exchange chromatography and treatment with activated carbon. Then, it is normally concentrated by evaporation to about 70% dry solids.

For many purposes a 42% fructose syrup is perfectly satisfactory for use but it does not match the exacting criteria of the quality soft drink manufacturers as a replacement for sucrose in acidic soft drinks. For use in the better colas, 55% fructose is required. This is produced by using vast chromatographic columns of zeolites or the calcium salts of cation exchange resins to adsorb and separate the fructose from the other components. The fractionation process, although basically very simple, is only economic if run continuously. The fructose stream (90% (w/w) fructose, 9% glucose) is blended with 42% fructose syrups to give the 55% fructose (42% glucose) product required. The glucose-rich 'raffinate' stream may be recycled but if this is done undesirable oligosaccharides build up in the system. Immobilised glucoamylase is used in some plants to hydrolyse oligosaccharides in the raffinate; here the substrate concentration is comparatively low (around 20% dry solids) so the formation of isomaltose by the enzyme is insignificant.

Clearly the need for a second large fructose enrichment plant in addition to the glucose isomerase plant is undesirable and attention is being paid to means of producing 55% fructose syrups using only the enzyme. The thermodynamics of the system favour fructose production at higher temperatures and 55% fructose syrups could be produced directly if the enzyme reactors were operated at around 95 °C. The use of miscible organic co-solvents may also produce the desired effect. Both these alternatives present a more than considerable challenge to enzyme technology!

The present world market for HFCS is over 5 million tonnes of which about 60% is for 55% fructose syrup with most of the remainder for 42% fructose syrup. This market is still expanding and ensures that HFCS production is the major application for immobilised-enzyme technology.

The high-fructose syrups can be used to replace sucrose where sucrose is used in solution but they are inadequate to replace crystalline sucrose.

Another ambition of the corn syrup industry is to produce sucrose from starch. This can be done using a combination of the enzymes phosphorylase (EC 2.4.1.1), glucose isomerase and sucrose phosphorylase (EC 2.4.1.7), but the thermodynamics do not favour the conversion so means must be found of removing sucrose from the system as soon as it is formed. This will not be easy but is achievable if the commercial pull (i.e. money available) is sufficient:

$$\text{starch (G}_n\text{)} + \text{orthophosphate} \underset{}{\overset{\text{phosphorylase}}{\rightleftharpoons}} \text{starch (G}_{n-1}\text{)} + \alpha\text{-glucose-1-phosphate} \quad [5.1]$$

$$\text{glucose} \underset{}{\overset{\text{glucose isomerase}}{\rightleftharpoons}} \text{fructose} \quad [5.2]$$

$$\alpha\text{-glucose-1-phosphate} + \text{fructose} \underset{}{\overset{\text{sucrose phosphorylase}}{\rightleftharpoons}} \text{sucrose} + \text{orthophosphate} \quad [5.3]$$

A further possible approach to producing sucrose from glucose is to supply glucose at high concentrations to microbes whose response to osmotic stress is to accumulate sucrose intracellularly. Provided they are able to release sucrose without hydrolysis when the stress is released, such microbes may be the basis of totally novel processes.

Use of immobilised raffinase

The development of a raffinase (α-D-galactosidase) suitable for commercial use is another triumph of enzyme technology. Plainly, it would be totally unacceptable to use an enzyme preparation containing invertase to remove this material during sucrose production (see Chapter 4). It has been necessary to find an organism capable of producing an α-galactosidase but not an invertase. A mould, *Mortierella vinacea* var. *raffinoseutilizer*, fills the requirements. This is grown in a particulate form and the particles harvested, dried and used directly as the immobilised-enzyme preparation. It is stirred with the sugar beet juice in batch stirred tank reactors. When the removal of raffinose is complete, stirring is stopped and the juice pumped off the settled bed of enzyme. Enzyme, lost by physical attrition, is replaced by new enzyme added with the next batch of juice. The galactose released is destroyed in the alkaline conditions of the first stages of juice purification and does not cause any further problems while the sucrose is recovered. This process results in a 3% increase in productivity and a significant reduction in the costs of the disposal of waste molasses.

Immobilised raffinase may also be used to remove the raffinose and stachyose from soybean milk. These sugars are responsible for the flatulence that may be caused when soybean milk is used as a milk substitute in special diets.

Use of immobilised invertase

Invertase was probably the first enzyme to be used on a large scale in an immobilised form (by Tate & Lyle). In the period 1941–1946 the acid, previously used in the manufacture of Golden Syrup, was unavailable, so yeast invertase was used instead. Yeast cells were autolysed and the autolysate clarified by adjustment to pH 4.7, followed by filtration through a bed of calcium sulphate and adsorption into bone char. A layer of the bone char containing invertase was included in the bed of bone char already used for decolourising the syrup. The scale used was large, the bed of invertase-char being 2 ft (60 cm) deep in a bed of char 20 ft (610 cm) deep. The preparation was very stable, the limiting factors being microbial contamination or loss of decolourising power rather than loss of enzymic activity. The process was cost-effective but, not surprisingly, the product did not have the subtlety of flavour of the acid-hydrolysed material and the immobilised enzyme process was abandoned when the acid became available once again. Recently, however, it has been relaunched using Brimac™, where the invertase–char mix is stabilised by cross-linking and has a half-life of 90 days in use (pH 5.5, 50 °C). The revival is due, in part, to the success of HFCS as a high-quality low-colour sweetener. It is impossible to produce inverted syrups of equivalent quality by acid hydrolysis. Enzymic inversion avoids the high-colour, high salt-ash, relatively low conversion and batch variability problems of acid hydrolysis. Although free invertase may be used (with residence times of about a day), the use of immobilised enzymes in a PBR (with residence time of about 15 min) makes the process competitive; the cost of 95% inversion (at 50% (w/w)) being no more than the final evaporation costs (to 75% (w/w)). A productivity of 16 tonnes of inverted syrup (dry weight) may be achieved using one litre of the granular enzyme.

Production of amino acids

Another early application of an immobilised enzyme was the use of the aminoacylase from *Aspergillus oryzae* to resolve racemic mixtures of amino acids.

$$
\underset{\substack{\text{N-acyl-DL-amino acid}}}{\text{DL } \underset{\substack{| \\ \text{NH}-\text{C}=\text{O} \\ | \\ \text{R}'}}{\overset{\substack{\text{O}^- \\ |}}{\text{RCHC}=\text{O}}}} + \text{H}_2\text{O} \xrightarrow{\text{aminoacylase}} \underset{\substack{\text{L-amino acid}}}{\text{L } \underset{\substack{| \\ ^+\text{NH}_3}}{\overset{\substack{\text{O}^- \\ |}}{\text{RCHC}=\text{O}}}} + \underset{\substack{\text{N-acyl-D-amino acid}}}{\text{D } \underset{\substack{/ \\ \text{NH}-\text{C}=\text{O} \\ | \\ \text{R}'}}{\overset{\substack{\text{O}^- \\ |}}{\text{RCHC}=\text{O}}}} + \underset{\substack{|}}{\overset{\substack{\text{O}^- \\ |}}{\text{R}'\text{C}=\text{O}}} \quad [5.4]
$$

Chemically synthesised racemic N-acyl-DL-amino acids are hydrolysed at pH 8.5 to give the free L-amino acids plus the unhydrolysed N-acyl-D-amino acids. These products are easily separated by differential crystallisation and the N-acyl-D-amino acids racemised chemically (or enzymically) and reprocessed. The enzyme is immobilised by adsorption to anion exchange resins (e.g. DEAE-Sephadex) and has an operational half-life of about 65 days at 50 °C in PBRs with residence times of about 30 min. The reactors may be re-activated *in situ* by simply adding more enzyme. The immobilised enzyme has proved a more economical process than the use of free enzyme mainly due to the more efficient use of the substrate and reductions in the cost of enzyme and labour.

Novel and natural L-amino acids can be produced by the chemical conversion of aldehydes through DL-amino nitriles to racemic DL-hydantoins (reaction scheme [5.5]) followed by enzymic hydrolysis with hydantoinase and a carbamoylase (reaction scheme [5.6]) at pH 8.5. Both enzymes may be obtained from *Arthrobacter* species.

D-Amino acids are important constituents in antibiotics and insecticides. They may be produced in a manner similar to the L-amino acids but using hydantoinases of differing specificity. The *Pseudomonas striata* enzyme is

$$
\text{RCH}=\text{O} + \text{HCN} + (\text{NH}_4)_2\text{CO}_3 \longrightarrow \underset{\substack{| \\ \text{NH}_2}}{\text{RCH}-\text{CN}} + \text{NH}_4\text{HCO}_3 + \text{H}_2\text{O}
$$

$$
\Big\downarrow \text{CO}_2
$$

$$
\underset{\substack{\text{DL-hydantoin}}}{\underset{\substack{| \quad | \\ \text{HN} \quad \text{NH} \\ \diagdown \text{C} \diagup \\ \| \\ \text{O}}}{\text{RCH}-\text{C}=\text{O}}} \qquad [5.5]
$$

$$
\underset{\substack{\text{DL-hydantoin}}}{\underset{\substack{| \quad | \\ \text{HN} \quad \text{NH} \\ \diagdown \text{C} \diagup \\ \| \\ \text{O}}}{\text{RCH}-\text{C}=\text{O}}} + \text{H}_2\text{O} \xrightarrow{\text{hydantoinase}} \underset{\substack{\text{DL-carbamoyl} \\ \text{amino acid}}}{\underset{\substack{| \\ \text{NH} \\ | \\ \text{C}=\text{O} \\ | \\ \text{NH}_2}}{\overset{\substack{\text{O}^- \\ |}}{\text{RCH}-\text{C}=\text{O}}}} \xrightarrow[\text{H}_2\text{O}]{\text{carbamoylase}} \underset{\substack{\text{L-amino acid}}}{\underset{\substack{| \\ \text{NH}_2}}{\overset{\substack{\text{O}^- \\ |}}{\text{RCH}-\text{C}=\text{O}}}} + \text{NH}_3 + \text{CO}_2 \quad [5.6]
$$

specific for D-hydantoins, allowing their specific hydrolysis to D-carbamoyl amino acids which can be converted to the D-amino acids by chemical treatment with nitrous acid. They remaining L-hydantoin may be simply racemised by base and the process repeated.

L-Aspartic acid is widely used in the food and pharmaceutical industries and is needed for the production of the low-calorific sweetener aspartame. It may be produced from fumaric acid by the use of the aspartate ammonia-lyase (aspartase) from *Escherichia coli*.

$$^-OOCCH{=}CHCOO^- + NH_4^+ \underset{}{\overset{\text{aspartate ammonia-lyase}}{\rightleftharpoons}} {^-OOCCH_2CHCOO^-} \quad [5.7]$$

$$\underset{+NH_3}{\overset{|}{}}$$

fumaric acid   L-aspartic acid

A crude immobilised aspartate ammonia-lyase ($50\,000\ U\ g^{-1}$) may be prepared by entrapping *Escherichia coli* cells in a κ-carageenan gel cross-linked with glutaraldehyde and hexamethylenediamine. The process is operated in a PBR at pH 8.5 using ammonium fumarate as the substrate, with a reported operational half-life of 680 days at 37 °C.

Urocanic acid is a sun-screening agent which may be produced from L-histidine by the histidine ammonia-lyase (histidase) from *Achromobacter liquidum* (see reaction scheme [1.4], p. 4). The organism cannot be used directly as it has urocanate hydratase activity, which removes the urocanic acid. However, a brief heat treatment (70 °C, 30 min) inactivates this unwanted activity but has little effect on the histidine ammonia-lyase. A crude immobilised-enzyme preparation consisting of heat-treated cells entrapped in a polyacrylamide gel has been used to effect this conversion, showing a half-life of 180 days at 37 °C.

Use of immobilised lactase

Lactase is one of relatively few enzymes that have been used both free and immobilised in large-scale processes. The reasons for its utility has been given earlier (see Chapter 4), but the relatively high cost of the enzyme is an added incentive for its use in an immobilised state.

Immobilised lactases are important mainly in the treatment of whey, as the fats and proteins in the milk emulsion tend to coat the biocatalysts. This both reduces their apparent activity and increases the probability of microbial colonisation.

Yeast lactase has been immobilised by incorporation into cellulose triace-tate fibres during wet spinning, a process developed by Snamprogetti S.p.A. in Italy. The fibres are cut up and used in a batchwise STR process at 5 °C

(*Kluyveromyces lactis*, pH optimum 6.4–6.8, 90 U g^{-1}). Fungal lactases have been immobilised on 0.5 mm diameter porous silica (35 nm mean pore diameter) using glutaraldehyde and γ-aminopropyltriethoxysilane (*Aspergillus niger*, pH optimum 3.0–3.5, 500 U g^{-1}; *A. oryzae*, pH optimum 4.0–4.5, 400 U g^{-1}). They are used in PBRs. Due to the different pH optima of fungal and yeast lactases, the yeast enzymes are useful at the neutral pH of both milk and sweet whey, whereas fungal enzymes are more useful with acid whey.

Immobilised lactases are particularly affected by two inherent shortcomings. Product inhibition by galactose and unwanted oligosaccharide formation are both noticeable under the diffusion-controlled conditions usually prevalent. Both problems may be reduced by an increase in the effectiveness factor and a reduction in the degree of hydrolysis or initial lactose concentration, but such conditions also lead to a reduction in the economic return. The control of microbial contamination within the bioreactors is the most critical practical problem in these processes. To some extent, this may be overcome by the use of regular sanitation with basic detergents and a dilute protease solution.

Production of antibiotics

Benzylpenicillins and phenoxymethylpenicillins (penicillins G and V, respectively) are produced by fermentation and are the basic precursors of a wide range of semi-synthetic antibiotics, e.g. ampicillin. The amide link may be hydrolysed conventionally but the conditions necessary for its specific hydrolysis, whilst causing no hydrolysis of the intrinsically more labile but pharmacologically essential β-lactam ring, are difficult to attain. Such specific hydrolysis may be simply achieved by use of penicillin amidases (also called penicillin acylases). Different enzyme preparations are generally used for the hydrolysis of the penicillins G and V, pencillin-V-amidase being much more specific than pencillin-G-amidase:

Penicillin amidase may be obtained from *E. coli* and has been immobilised on a number of supports including cyanogen bromide-activated Sephadex G200. It represents one of the earliest successful processes involving immobilised enzymes and is generally used in batch or semicontinuous STR

$$\phi-CH_2-\overset{\overset{O}{\|}}{C}-N-\overset{H}{\underset{\|}{C}}-\overset{H}{\underset{\|}{C}}-\overset{H}{\overset{S}{C}}C(CH_3)_2 \quad + H_2O \longrightarrow \quad H_3N^+-\overset{H}{\underset{\|}{C}}-\overset{H}{\overset{S}{C}}C(CH_3)_2 \quad + \phi-CH_2-CO_2^- \quad [5.8]$$

benzyl penicillin   6-aminopenicillanic acid
+ phenylacetic acid

$$\phi - O - CH_2 - \overset{\overset{O}{\|}}{C} - N - \overset{\overset{H}{|}}{C} - \overset{\overset{H}{|}}{C} - \overset{\overset{H}{|}}{C}\overset{S}{\diagdown}C(CH_3)_2 \atop \underset{\underset{H}{|}}{O = C - N - C - CO_2^-} + H_2O \longrightarrow \quad H_3\overset{+}{N} - \overset{\overset{H}{|}}{C} - \overset{\overset{H}{|}}{C}\overset{S}{\diagdown}C(CH_3)_2 \atop \underset{\underset{H}{|}}{O = C - N - C - CO_2^-} + \phi - O - CH_2 - CO_2^- \quad [5.9]$$

phenoxymethyl penicillin   6-aminopenicillanic acid + phenoxyacetic acid

processes ($40\,000\,U\,kg^{-1}$ penicillin G, 35 °C, pH 7.8, 2 h) where it may be reused over 100 times. It has also been used in PBRs, where it has an active life of over 100 days, producing about 2 tonnes of 6-aminopenicillanic acid kg^{-1} of immobolised enzyme.

The penicillin-G-amidases may be used 'in reverse' to synthesise penicillin and cephalosporin antibiotics by non-equilibrium kinetically controlled reactions (see also Chapter 7). Ampicillin has been produced by the use of penicillin-G-amidase immobilised by adsorption to DEAE-cellulose in a packed bed column:

$$H_3\overset{+}{N} - \overset{\overset{H}{|}}{C} - \overset{\overset{H}{|}}{C}\overset{S}{\diagdown}C(CH_3)_2 \atop \underset{\underset{H}{|}}{O = C - N - C - CO_2^-} + \underset{\underset{NH_3^+}{|}}{\phi - CH - CO_2CH_3} \longrightarrow \quad \underset{\underset{NH_3^+}{|}}{\phi - CH} - \overset{\overset{O}{\|}}{C} - N - \overset{\overset{H}{|}}{C} - \overset{\overset{H}{|}}{C}\overset{S}{\diagdown}C(CH_3)_2 \atop \underset{\underset{H}{|}}{O = C - N - C - CO_2^-} + CH_3OH \quad [5.10]$$

6-aminopenicillanic acid   ampicillin   + methanol
+ D-phenylglycine methyl ester

Many other potential and proven antibiotics have been synthesised in this manner, using a variety of synthetic β-lactams and activated carboxylic acids.

Preparation of acrylamide

Acrylamide is an important monomer needed for the production of a range of economically useful polymeric materials. It may be produced by the addition of water to acrylonitrile.

$$CH_2{=}CHCN + H_2O \longrightarrow CH_2{=}CHCONH_2 \qquad [5.11]$$

This process may be achieved by the use of a reduced copper catalyst (Cu^+); however, the yield is poor, unwanted polymerisation or conversion to acrylic acid ($CH_2{=}CHCOOH$) may occur at the relatively high temperatures involved (80–140 °C) and the catalyst is difficult to regenerate. These problems may be overcome by the use of immobilised nitrile hydratase (often erroneously called a nitrilase). The enzyme from *Rhodococcus* has been used by the Nitto Chemical Industry Co. Ltd, as it contains only very low amidase activity which otherwise would produce unwanted acrylic acid from the acrylamide.

Immobilised nitrile hydratase is simply prepared by entrapping the intact cells in a cross-linked 10% (w/v) polyacrylamide/dimethylaminoethylmethacrylate gel and granulating the product. It is used at 10 °C and pH 8.0–8.5 in a semibatchwise process, keeping the substrate acrylonitrile concentration below 3% (w/v). Using 1% (w/v) immobilised-enzyme concentrations (about $50\,000\,Ul^{-1}$) the process takes about a day. Product concentrations of up to 20% (w/v) acrylamide have been achieved, containing negligible substrate and less than 0.02% (w/w) acrylic acid. Acrylamide production using this method is about 4000 tonnes per year.

The closely related enzymes cyanidase and cyanide hydratase (see schemes [5.12] and [5.13], respectively) are used to remove cyanide from industrial waste and in the detoxification of feeds and foodstuffs containing amygdalin (see equation [6.12], p. 207).

$$HCN + 2H_2O \longrightarrow HCOO^- + NH_4^+ \qquad [5.12]$$

$$HCN + H_2O \longrightarrow HCONH_2 \qquad [5.13]$$

Summary

(*a*) Enzymes may be used in batch or continuous reactors. The choice of reactor for specific applications depends upon the physical forms of the substrate(s), enzyme(s) and product(s) in addition to the process economics.

(*b*) The PBR generally makes the most efficient use of immobilised enzymes. The CSTR is preferred mainly where pH control is essential or where the physical properties of the substrate stream are incompatible with PBR operation. The FBR has properties between those of a PBR and a CSTR but is more difficult to scale-up. Membrane reactors are versatile but expensive.

(*c*) Immobilised enzymes are used in an increasing number of economically viable bioconversions, the most important of which is the isomerisation of glucose.

Bibliography

Baret, J. L. (1987). Large-scale production and application of immobilised lactase. *Methods in Enzymology*, **136**, 411–23.

Chen, W.-P. (1980). Glucose isomerase (a review). *Process Biochemistry*, **15**, 30–41.

Daniels, M. J. (1986). Immobilised enzymes in carbohydrate food processing. In *Biotechnology in the food industry*, pp. 29–35. Pinner: Online Publications.

Daniels, M. J. (1987). Industrial operation of immobilised enzymes. *Methods in Enzymology*, **136**, 356–70.

Howaldt, M. W., Kulbe, K. D. & Chmiel, H. (1986). Choice of reactor to minimise enzyme requirement. 1. Mathematical model for one-substrate Michaelis–Menten type kinetics in continuous reactors. *Enzyme and Microbial Technology*, **8**, 627–31.

Jensen, V. J. & Rugh, S. (1987). Industrial-scale production and application of immobilised glucose isomerase. *Methods in Enzymology*, **136**, 356–70.

Kleinstreuer, C. & Agarwal, S. S. (1986). Analysis and simulation of hollow-fiber bioreactor dynamics. *Biotechnology and Bioengineering*, **28**, 1233–40.

Soda, K. & Yonaha, K. (1987). Application of free enzymes in pharmaceutical and chemical industries. In *Biotechnology*, vol. 7a *Enzyme technology*, ed. J. F. Kennedy, pp. 605–52. Weinheim: VCH Verlagsgesellschaft mbH.

Van Tilburg (1984). Enzymic isomerisation of corn starch-based glucose syrups. In *Starch conversion technology*, ed. G. M. A. Van Beynum & J. A. Roels, pp. 175–236. New York: Marcel Dekker Inc.

Vieth, W. R., Venkatasubramanian, K., Constantinides, A. & Davidson, B. (1976). Design and analysis of immobilized-enzyme flow reactors. In *Applied biochemistry and bioengineering*, vol. 1 *Immobilised enzyme principles*, ed. L. B. Wingard & E. Katchalski-Katzir, pp. 221–327. London: Academic Press.

Watanabe, I. (1987). Acrylamide production method using immobilised nitrilase-containing microbial cells. *Methods in Enzymology*, **136**, 523–30.

6   Biosensors

The use of enzymes in analysis

Enzymes make excellent analytical reagents, due to their specificity, selectivity and efficiency. They are often used to determine the concentration of their substrates (as analytes) by means of the resultant initial reaction rates. If the reaction conditions and enzyme concentrations are kept constant, these rates of reaction (v) are proportional to the substrate concentrations ([S]) at low substrate concentrations. When [S] < 0.1 K_m, equation (1.8) (p. 9) simplifies to give:

$$v = (V_{max}/K_m)[S] \qquad (6.1)$$

The rates of reaction are commonly determined from the difference in optical absorbance between the reactants and products. An example of this is the β-D-galactose dehydrogenase (EC 1.1.1.48) assay for galactose which involves the oxidation of galactose by the redox coenzyme, nicotine-adenine dinucleotide (NAD^+):

$$\text{β-D-galactose} + NAD^+ \rightarrow \text{D-galactono-1,4-lactone} + NADH + H^+ \qquad [6.1]$$

A 0.1 mM solution of NADH has an absorbance of 0.622 at 340 nm in a 1 cm path-length cuvette, whereas the NAD^+ from which it is derived has effectively zero absorbance at this wavelength. The conversion ($NAD^+ \rightarrow NADH$) is, therefore, accompanied by a large increase in absorption of light at this wavelength. For the reaction to be linear with respect to the galactose concentration, the galactose is kept within a concentration range well below the K_m of the enzyme for galactose. In contrast, the NAD^+ concentration is kept within a concentration range well above the K_m of the enzyme for NAD^+, in order to avoid limiting the reaction rate. Such assays are commonly used in analytical laboratories and are, indeed, excellent where a wide variety of analyses need to be undertaken on a relataively small number of samples. The drawbacks to this type of analysis become apparent when a large number of repetitive assays need to be performed. Then, they are seen to be costly in terms of expensive enzyme and coenzyme usage, time consuming, labour intensive and in need of skilled and reproducible operation within properly equipped analytical laboratories. For routine or on-site operation, these disadvantages must be overcome. This is being achieved by the production of biosensors which exploit biological systems in association with advances in micro-electronic technology.

What are biosensors?

A biosensor is an analytical device which converts a biological response into an electrical signal (Figure 6.1). The term biosensor is often used to cover sensor devices used in order to determine the concentration of substances and other parameters of biological interest even where they do not utilise a biological system directly. This very broad definition is used by some scientific journals (e.g. *Biosensors*, Elsevier Applied Science) but will not be applied to the coverage here. The emphasis of this chapter concerns enzymes as the biologically responsive material, but it should be recognised that other biological systems may be utilised by biosensors. e.g. whole cell metabolism, ligand binding and the antibody–antigen reaction. Biosensors represent a rapidly expanding field at the present time, with an estimated 60% annual growth rate; the major impetus comes from the health-care industry (e.g. 6% of the people of the Western world are diabetic and would benefit from the availability of a rapid, accurate and simple biosensor for glucose) but there is some pressure from other areas, such as food quality appraisal and environmental monitoring. The estimated world analytical market is about £12 000 000 000 year^{-1}, of which 30% is in the health care area. There is clearly a vast market expansion potential as less than 0.1% of this market is currently using biosensors. Research and development in this field is wide and multidisciplinary, spanning biochemistry, bioreactor science, physical chemistry, electrochemistry, electronics and software engineering. Most of this current endeavour concerns potentiometric and amperometric biosensors and colorimetric paper enzyme strips. However, all the main transducer types are likely to be thoroughly examined, for use in biosensors, over the next few years.

A successful biosensor must possess at least some of the following beneficial features.

(1) The biocatalyst must be highly specific for the purpose of the analyses, be stable under normal storage conditions and, except in the case of colorimetric enzyme strips and dipsticks (see later), show good stability over a large number of assays (i.e. much greater than 100).

(2) The reaction should be as independent of such physical parameters as stirring, pH and temperature as is manageable. This would allow

the analysis of samples with minimal pre-treatment. If the reaction involves cofactors or coenzymes these should, preferably, also be co-immobilised with the enzyme (see Chapter 8).

(3) The response should be accurate, precise, reproducible and linear over the useful analytical range, without dilution or concentration. It should also be free from electrical noise.

(4) If the biosensor is to be used for invasive monitoring in clinical situations, the probe must be tiny and biocompatible, having no toxic or antigenic effects. If it is to be used in fermenters, it should be sterilisable. This is preferably performed by autoclaving but no biosensor enzymes can presently withstand such drastic wet-heat treatment. In either case, the biosensor should not be prone to fouling or proteolysis.

(5) The complete biosensor should be cheap, small, portable and capable of being used by semi-skilled operators.

(6) There should be a market for the biosensor. There is clearly little purpose in developing a biosensor if other factors (e.g. government subsidies, the continued employment of skilled analysts, or poor customer perception) encourage the use of traditional methods and discourage the decentralisation of laboratory testing.

The biological response of the biosensor is determined by the biocatalytic membrane which accomplishes the conversion of reactant to product. Immobilised enzymes possess a number of advantageous features which make them particularly applicable for use in such systems. They may be reused, which ensures that the same catalytic activity is present for a series of analyses. This is an important factor in securing reproducible results and avoids the pitfalls associated with the replicate pipetting of free enzyme otherwise necessary in analytical protocols. Many enzymes are intrinsically stabilised by the immobilisation process (see Chapter 3), but even where this does not occur there is usually considerable apparent stabilisation. It is normal to use an excess of the enzyme within the immobilised sensor system. This gives a catalytic redundancy (i.e. $\eta \ll 1$) which is sufficient to ensure an increase in the apparent stabilisation of the immobilised enzyme (see, for example, Figures 3.11 (p. 105), 3.19 (p. 118) and 5.8 (p. 181). Even where there is some inactivation of the immobilised enzyme over a period of time, this inactivation is usually steady and predictable. Any activity decay is easily incorporated into an analytical scheme by regularly interpolating standards between the analyses of unknown samples. For these reasons, many such immobilised-enzyme systems are reusable up to 10 000 times over a period of several months. Clearly, this results in a considerable saving in terms of the cost of the enzymes relative to the analytical usage of free soluble enzymes.

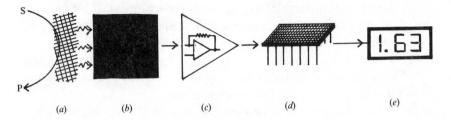

Figure 6.1. Diagram showing the main components of a biosensor. The biocatalyst (*a*) converts the substrate to product. This reaction is determined by the transducer (*b*), which converts it to an electrical signal. The output from the transducer is amplified (*c*), processed (*d*) and displayed (*e*).

When the reaction occurring at the immobilised-enzyme membrane of a biosensor is limited by the rate of external diffusion, the reaction process will possess a number of valuable analytical assets. In particular, it will obey the relationship shown in equation (3.27) (p. 106). It follows that the biocatalyst gives a proportional change in reaction rate in response to the reactant (substrate) concentration over a substantial linear range, several times the intrinsic K_m (see Figure 3.12, line *e*, p. 107). This is very useful as analyte concentrations are often approximately equal to the K_m values of their appropriate enzymes, i.e. roughly 10 times more concentrated than can be normally determined, without dilution, by use of the free enzymes in solution. Also following from equation (3.27) is the independence of the reaction rate with respect to pH, ionic strength, temperature and inhibitors. This simply avoids the tricky problems often encountered due to the variability of real analytical samples (e.g. fermentation broth, blood and urine) and external conditions. Control of biosensor response by the external diffusion of the analyte can be encouraged by the use of permeable membranes between the enzyme and the bulk solution. The thickness of these can be varied with associated effects on the proportionality constant between the substrate concentration and the rate of reaction (i.e. increasing membrane thickness increases the unstirred layer (δ) which, in turn, decreases the proportionality constant, K_L, in equation (3.27)). Even if total dependence on the external diffusional rate is not achieved (or achievable), any increase in the dependence of the reaction rate on external or internal diffusion will cause a reduction in the dependence on the pH, ionic strength, temperature and inhibitor concentrations.

The key part of the biosensor is the transducer (shown as the 'black box' in Figure 6.1) which makes use of a physical change accompanying the reaction. This may be (1) the heat output (or absorbed) by the reaction (calorimetric

Table 6.1 *Heat output (molar enthalpies) of enzyme-catalysed reactions*

Reactant	Enzyme	Heat output $-\Delta H$ (kJ mol^{-1})
Cholesterol	Cholesterol oxidase	53
Esters	Chymotrypsin	4–16
Glucose	Glucose oxidase	80
Hydrogen peroxide	Catalase	100
Penicillin G	Penicillinase	67
Peptides	Trypsin	10–30
Starch	Amylase	8
Sucrose	Invertase	20
Urea	Urease	61
Uric acid	Uricase	49

biosensors), (2) changes in the distribution of charges causing an electrical potential to be produced (potentiometric biosensors), (3) movement of electrons produced in a redox reaction (amperometric biosensors), (4) light output during the reaction or a light absorbance difference between the reactants and products (optical biosensors), or (5) effects due to the mass of the reactants or products (piezo-electric biosensors).

There are three so-called generations of biosensors: first-generation biosensors, where the normal product of the reaction diffuses to the transducer and causes the electrical response; second-generation biosensors, which involve specific 'mediators' between the reaction and the transducer in order to generate improved response; and third-generation biosensors, where the reaction itself causes the response and no product or mediator diffusion is directly involved.

The electrical signal from the transducer is often low and superimposed upon a relatively high and noisy baseline (i.e. a high-frequency signal component of an apparently random nature, due to electrical interference or generated within the electronic components of the transducer). The signal processing normally involves subtracting a 'reference' baseline signal, derived from a similar transducer without any biocatalytic membrane, from the sample signal, amplifying the resultant signal difference and electronically filtering (smoothing) out the unwanted signal noise. The relatively slow nature of the biosensor response considerably eases the problem of electrical noise filtration. The analogue signal produced at this stage may be output directly but is usually converted to a digital signal and passed to a microprocessor stage where the data is processed, converted to a concentration units and output to a display device or data store.

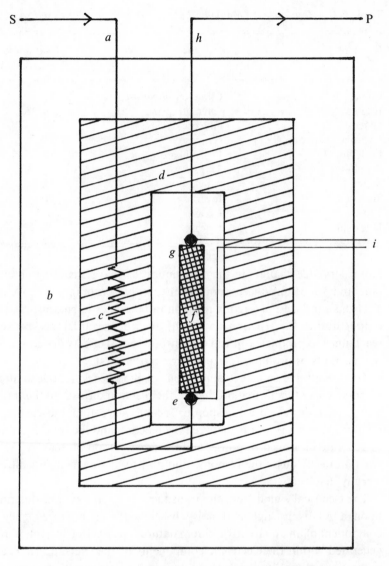

Figure 6.2. Diagram of a calorimetric biosensor. The sample stream (*a*)
passes through the outer insulated box (*b*) to the heat exchanger (*c*) within
an aluminium block (*d*). From there, it flows past the reference thermistor
(*e*) and into the packed bed bioreactor (*f*, 1 ml volume), containing the
biocatalyst, where the reaction occurs. The change in temperature is
determined by the thermistor (*g*) and the solution passed to waste (*h*).
External electronics (*i*) determine the difference in the resistance, and
hence temperature, between the thermistors.

Calorimetric biosensors

Many enzyme-catalysed reactions are exothermic, generating heat (Table 6.1) which may be used as a basis for measuring the rate of reaction and, hence, the analyte concentration. This represents the most generally applicable type of biosensor. The temperature changes are usually determined by means of thermistors at the entrance and exit of small packed bed columns containing immobilised enzymes within a constant-temperature environment (Figure 6.2). Under such closely controlled conditions, up to 80% of the heat generated in the reaction may be registered as a temperature change in the sample stream. This may be simply calculated from the enthalpy change and the amount reacted. If a 1 mM reactant is completely converted to product in a reaction generating 100 kJ mol^{-1} then each millilitre of solution generates 0.1 J of heat. At 80% efficiency, this will cause a change in temperature of the solution amounting to approximately 0.02 deg. C. This is about the temperature change commonly encountered and necessitates a temperature resolution of 0.0001 deg. C for the biosensor to be generally useful.

The thermistors, used to detect the temperature change, function by changing their electrical resistance with the temperature, obeying the relationship:

$$\ln(R_1/R_2) = B\{(1/T_1) - (1/T_2)\} \tag{6.2}$$

where R_1 and R_2 are the resistances of the thermistors at absolute temperatures T_1 and T_2, respectively, and B is a characteristic temperature constant for the thermistor. When the temperature change is very small, as in the present case, $B\{(1/T_1) - (1/T_2)\}$ is very much smaller than 1 and this relationship may be substantially simplified using the approximation when $x \ll 1$ that $e^x \approx 1 + x$ (x here being $B\{(1/T_1) - (1/T_2)\}$).
Therefore:

$$R_1 = R_2\{1 + B(T_2 - T_1)/T_1T_2\} \tag{6.3}$$

As $T_1 \approx T_2$, they may both be replaced in the denominator by T_1.
Therefore:

$$\Delta R/R = -(B/T_1^2)\Delta T \tag{6.4}$$

The relative decrease in the electrical resistance ($\Delta R/R$) of the thermistor is proportional to the increase in temperature (ΔT). A typical proportionality constant ($-B/T_1^2$) is -4% deg. C^{-1}. The resistance change is converted to a proportional voltage change, using a balanced Wheatstone bridge incorporating precision wire-wound resistors, before amplification. The expectation that there will be a linear correlation between the response and the enzyme activity has been found to be borne out in practice. A major problem with this biosensor is the difficulty encountered in closely matching the character-istic temperature constants of the measurement and reference thermistors. An equal movement of only 1 deg. C in the background temperature of both thermistors commonly causes an apparent change in the relative resistances of the thermistors equivalent to 0.01 deg. C and equal to the full-scale change due to the reaction. It is clearly of great importance that such environmental temperature changes are avoided, which accounts for inclusion of the well-insulated aluminium block in the biosensor design (see Figure 6.2).

The sensitivity (10^{-4} M) and range (10^{-4}–10^{-2} M) of thermistor biosensors are both quite low for the majority of applications, although greater sensitivity is possible using the more exothermic reactions (e.g. catalase). The low sensitivity of the system can be increased substantially by increasing the heat output by the reaction. In the simplest case this can be achieved by linking together several reactions in a reaction pathway, all of which contribute to the heat output. Thus the sensitivity of the glucose analysis using glucose oxidase can be more than doubled by the co-immobilisation of catalase within the column reactor in order to utilise the hydrogen peroxide produced. An extreme case of this amplification is shown in the following recycle scheme for the detection of ADP:

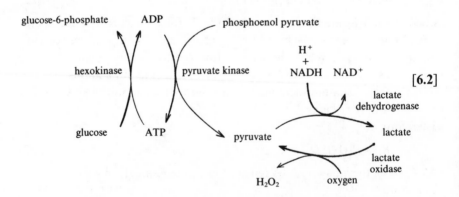

$$[6.2]$$

ADP is the added analyte and excess glucose, phosphoenol pyruvate, NADH and oxygen are present to ensure maximum reaction. Four enzymes (hexokinase, pyruvate kinase, lactate dehydrogenase and lactate oxidase) are co-immobilised within the packed bed reactor. In spite of the positive enthalpy of the pyruvate kinase reaction, the overall process results in a 1000-fold increase in sensitivity, primarily due to the recycling between pyruvate and lactate. Reaction limitation due to low oxygen solubility may be overcome by replacing it with benzoquinone, which is reduced to

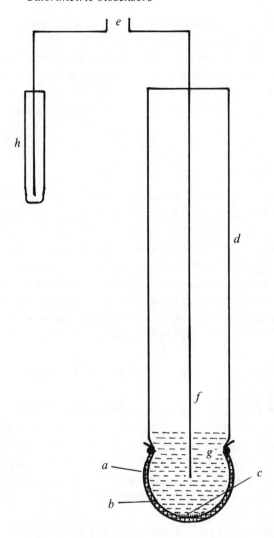

Figure 6.3. A simple potentiometric biosensor. A semipermeable membrane (*a*) surrounds the biocatalyst (*b*) entrapped next to the active glass membrane (*c*) of a pH probe (*d*). The electrical potential (*e*) is generated between the internal Ag/AgCl electrode (*f*) bathed in dilute HCl (*g*) and an external reference electrode (*h*).

hydroquinone by flavoenzymes. Such reaction systems do, however, have a serious disadvantage in that they increase the probability of the occurrence of interference in the determination of the analyte of interest. Reactions involving the generation of hydrogen ions can be made more sensitive by the inclusion of a base having a high heat of protonation. For example, the heat output by the penicillinase reaction may be almost doubled by the use of Tris (2-amino-2-hydroxymethylpropane-1, 3-diol) as the buffer.

In conclusion, the main advantages of the thermistor biosensor are its general applicability and the possibility for its use on turbid or strongly coloured solutions. The most important disadvantage is the difficulty in ensuring that the temperature of the sample stream remains constant ($\pm$ 0.01 deg. C).

Potentiometric biosensors

Potentiometric biosensors make use of ion-selective electrodes in order to transduce the biological reaction into an electrical signal. In the simplest terms this consists of a membrane, containing immobilised enzyme, surrounding the probe from a pH meter (Figure 6.3), where the catalysed reaction generates or absorbs hydrogen ions (Table 6.2). The reaction occurring next to the thin sensing glass membrane causes a change in pH which may be read directly from the pH meter's display. Typical of the use of such electrodes is that the electrical potential is determined at very high impedance, allowing effectively zero current flow and causing no interference with the reaction.

There are three types of ion-selective electrodes which are of use in biosensors.

(1) Glass electrodes for cations (e.g. normal pH electrodes) in which the sensing element is a very thin hydrated glass membrane which generates a transverse electrical potential due to the concentration-dependent competition between the cations for specific binding sites. The selectivity of this membrane is determined by the composition of the glass. The sensitivity to H^+ is greater than that achievable for NH_4^+.

(2) Glass pH electrodes coated with a gas-permeable membrane selective for CO_2, NH_3 or H_2S. The diffusion of the gas through this membrane causes a change in pH of a sensing solution between the membrane and the electrode which is then determined.

(3) Solid-state electrodes where the glass membrane is replaced by a thin membrane of a specific ion conductor made from a mixture of silver sulphide and a silver halide. The iodide electrode is useful for the determination of I^- in the peroxidase reaction (Table 6.2(c)) and also responds to cyanide ions.

The response of an ion-selective electrode is given by:

$$E = E_o + (RT/zF)\ln[i] \tag{6.5}$$

Table 6.2 *Reactions involving the release or absorption of ions that may be utilised by potentiometric biosensors*

(*a*) H^+ cation

$$\text{D-glucose} + O_2 \xrightarrow{\text{glucose oxidase}} \text{D-glucono-1,5-lactone} \xrightarrow{H_2O} \text{D-gluconate} + H^+ \qquad [6.3]$$
$$+$$
$$H_2O_2$$

$$\text{penicillin} \xrightarrow{\text{penicillinase}} \text{penicilloic acid} + H^+ \qquad [6.4]$$

$$H_2NCONH_2 + H_2O + 2H^+ \xrightarrow{\text{urease (pH 6.0)}^a} 2NH_4^+ + CO_2 \qquad [6.5]$$

$$H_2NCONH_2 + 2H_2O \xrightarrow{\text{urease (pH 9.5)}^b} 2NH_3 + HCO_3^- + H^+ \qquad [6.6]$$

$$\text{neutral lipids} + H_2O \xrightarrow{\text{lipase}} \text{glycerol} + \text{fatty acids} + H^+ \qquad [6.7]$$

(*b*) NH_4^+ cation

$$\text{L-amino acid} + O_2 + H_2O \xrightarrow{\text{L-amino acid oxidase}} \text{keto acid} + NH_4^+ + H_2O_2 \qquad [6.8]$$

$$\text{L-asparagine} + H_2O \xrightarrow{\text{asparaginase}} \text{L-aspartate} + NH_4^+ \qquad [6.9]$$

$$H_2NCONH_2 + 2H_2O + H^+ \xrightarrow{\text{urease (pH 7.5)}} 2NH_4^+ + HCO_3^- \qquad [6.10]$$

(*c*) I^- anion

$$H_2O_2 + 2H^+ + 2I^- \xrightarrow{\text{peroxidase}} I_2 + 2H_2O \qquad [6.11]$$

(*d*) CN^- anion

$$\text{amygdalin} + 2H_2O \xrightarrow{\beta\text{-glucosidase}} 2 \text{ glucose} + \text{benzaldehyde} + H^+ + CN^- \qquad [6.12]$$

[a] Can also be used in NH_4^+ and CO_2 (gas) potentiometric biosensors.
[b] Can also be used in an NH_3 (gas) potentiometric biosensor.

where E is the measured potential (in volts), E_o is a characteristic constant for the ion-selective/external electrode system, R is the gas constant, T is the absolute temperature (K), z is the signed ionic charge, F is the Faraday, and [i] is the concentration of the free uncomplexed ionic species (strictly, [i] should be the activity of the ion but at the concentrations normally encountered in biosensors, this is effectively equal to the concentration). This means, for example, that there is an increase in the electrical potential of 59 mV for every

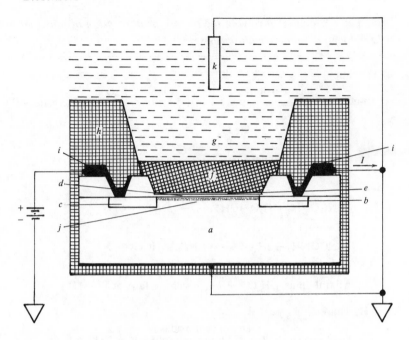

Figure 6.4. Diagram of the section across the width of an ENFET. The
actual dimensions of the active area are about 500 μm long by 50 μm
wide by 300 μm thick. The main body of the biosensor is a p-type silicon
chip (*a*) with two n-type silicon areas; the negative source (*b*) and the
positive drain (*c*). The chip is insulated by a thin layer (0.1 μm thick) of
silica (SiO_2, *d*) which forms the gate of the FET. Above this gate is an
equally thin layer of H^+-sensitive material (*e*, e.g. tantalum oxide), the
biocatalyst (*f*) and the analyte solution (*g*), which is separated from
sensitive parts of the FET by an inert encapsulating polyimide
photopolymer (*h*). When a potential is applied between the electrodes (*i*),
a current flows through the FET dependent upon the positive potential
detected at the ion-selective gate and its consequent attraction of electrons
into the depletion layer (*j*). This current (*I*) is compared with that from a
similar, but non-catalytic ISFET (*k*) immersed in the same solution. (Note
that the electric current is, by convention, in the opposite direction to the
flow of electrons.)

order of magnitude increase in the concentration of H^+ at 25 °C. The
logarithmic dependence of the potential on the ionic concentration is
responsible both for the wide analytical range and the low accuracy and
precision of these sensors. Their normal range of detection is 10^{-4}–10^{-2}M,
although a minority are 10-fold more sensitive. Typical response times are
between 1 and 5 min, allowing up to 30 analyses every hour.

Biosensors which involve H^+ release or utilisation necessitate the use of
very weakly buffered solutions (i.e. < 5 mM) if a significant change in

potential is to be determined. The relationship between pH change and substrate concentration is complex, including other such non-linear effects as pH–activity variation and protein buffering. However, conditions can often be found where there is a linear relationship between the apparent change in pH and the substrate concentration.

A recent development from ion-selective electrodes is the production of *ion-selective field effect transistors* (ISFETs) and their biosensor use as *enzyme-linked field effect transistors* (ENFETs, Figure 6.4). Enzyme membranes are coated on the ion-selective gates of these electronic devices, the biosensor responding to the electrical potential change via the current output. Thus, these are potentiometric devices, although they produce changes directly in the electric current. The main advantage of such devices is their extremely small size ($\ll 0.1$ mm^2), which allows cheap mass-produced fabrication using integrated circuit technology. As an example, a FET (ENFET containing bound urease with a reference electrode containing bound glycine) has been shown to show only a 15% variation in response to urea (0.05–10.0 mg ml^{-1}) during its active lifetime of a month. Several analytes may be determined by miniaturised biosensors containing arrays of ISFETs and ENFETs. The sensitivity of FETs, however, may be affected by the composition, ionic strength and concentrations of the solutions analysed.

Amperometric biosensors

Amperometric biosensors function by the production of a current when a potential is applied between two electrodes. They generally have response times, dynamic ranges and sensitivities similar to the potentiometric biosensors. The simplest amperometric biosensors in common usage involve the Clark oxygen electrode (Figure 6.5). This consists of a platinum cathode at which oxygen is reduced and a silver/silver chloride reference electrode. When a potential of -0.6 V, relative to the Ag/AgCl electrode is applied to the platinum cathode, a current proportional to the oxygen concentration is produced. Normally both electrodes are bathed in a solution of saturated potassium chloride and separated from the bulk solution by an oxygen-permeable plastic membrane (e.g. Teflon, polytetrafluoroethylene). The following reactions occur:

$$\text{Ag anode} \quad 4Ag + 4Cl^- \rightarrow 4AgCl + 4e^- \qquad [6.13]$$
$$\text{Pt cathode} \quad O_2 + 4H^+ + 4e^- \rightarrow 2H_2O \qquad [6.14]$$

The efficient reduction of oxygen at the surface of the cathode causes the oxygen concentration there to be effectively zero. The rate of this electrochemical reduction therefore depends on the rate of diffusion of the oxygen from

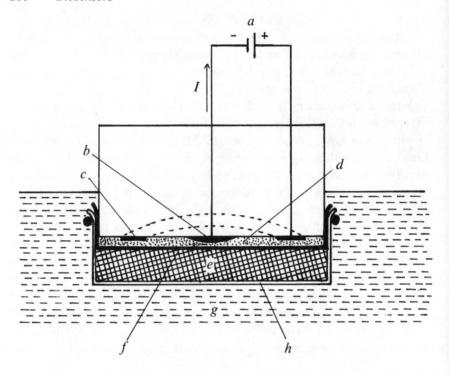

Figure 6.5. Diagram of a simple amperometric biosensor. A potential (*a*) is applied between the platinum cathode (*b*) and the annular silver anode (*c*). This generates a current (*I*) which is carried between the electrodes by means of a saturated solution of KCl (*d*). This electrode compartment is separated from the biocatalyst (*e*) by a thin plastic membrane (*f*), permeable only to oxygen. The analyte solution (*g*) is separated from the biocatalyst by another membrane (*h*), permeable to the substrate(s) and product(s). This biosensor is normally about 1 cm in diameter but has been scaled down to 0.25 mm diameter using a Pt wire cathode within a silver-plated steel needle anode and utilising dip-coated membranes.

the bulk solution, which is dependent on the concentration gradient and hence the bulk oxygen concentration (see, for example, equation (3.13) , p. 103). It is clear that a small, but significant, proportion of the oxygen present in the bulk is consumed by this process; the oxygen electrode measures the rate of a process which is far from equilibrium, whereas ion-selective electrodes are used close to equilibrium conditions. This causes the oxygen electrode to be much more sensitive to changes in the temperature than potentiometric sensors. A typical application for this simple type of biosensor is the determination of glucose concentrations by the use of a membrane containing immobilised glucose oxidase. The reaction (see reac-

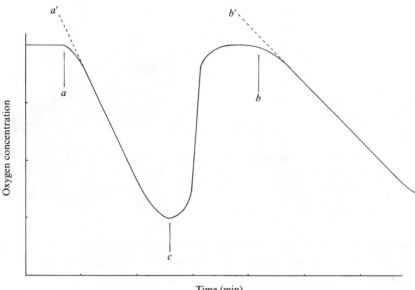

Figure 6.6. The response of an amperometric biosensor utilising glucose oxidase to the presence of 2mM (*a*) and 1 mM (*b*) glucose solutions. Between these analyses the biosensor was placed in oxygenated buffer (*c*) devoid of glucose. The steady rates of oxygen depletion (*a'* and *b'*) may be used to generate standard response curves and determine unknown samples. The time required for an assay can be considerably reduced if only the initial transient (curved) part of the response need be used, via a suitable model and software. The wash-out time, which roughly equals the time the electrode spends in the sample solution, is also reduced significantly by this process.

tion scheme [**1.1**], p. 3) results in a reduction of the oxygen concentration as it diffuses through the biocatalytic membrane to the cathode, this being detected by a reduction in the current between the electrodes (Figure 6.6). Other oxidases may be used in a similar manner for the analysis of the substrates (e.g. alcohol oxidase, D- and L-amino acid oxidases, cholesterol oxidase, galactose oxidase and uricase).

An alternative method for determining the rate of this reaction is to measure the production of hydrogen peroxide directly by applying a potential of $+0.68$ V to the platinum electrode, relative to the Ag/AgCl electrode, and causing the reactions:

$$\text{Pt anode} \quad H_2O_2 \rightarrow O_2 + 2H^+ + 2e^- \qquad [6.15]$$
$$\text{Ag cathode} \quad 2AgCl + 2e^- \rightarrow 2Ag + 2Cl^- \qquad [6.16]$$

A major problem with these biosensors is their dependence on the dissolved oxygen concentration. This may be overcome by the use of

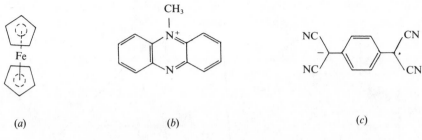

(a)   (b)   (c)

Figure 6.7. (*a*) Ferrocene (η^5-bis-cyclopentadienyl iron), the parent compound of a number of mediators. (*b*) NMP$^+$, the cationic part of conducting organic crystals. (*c*) TCNQ$^{\bar{}}$, the anionic part of conducting organic crystals. It is a resonance-stabilised radical formed by the one electron oxidation of TCNQH$_2$.

'mediators', which transfer the electrons directly to the electrode, bypassing the reduction of the oxygen co-substrate. In order to be generally applicable these mediators must possess a number of useful properties.

(1) They must react rapidly with the reduced form of the enzyme.

(2) They must be sufficiently soluble, in both the oxidised and reduced forms, to be able to diffuse rapidly between the active site of the enzyme and the electrode surface. This solubility should, however, not be so great as to cause significant loss of the mediator from the biosensor's microenvironment to the bulk of the solution. However soluble, the mediator should generally be non-toxic.

(3) The overpotential for the regeneration of the oxidised mediator, at the electrode, should be low and independent of pH.

(4) The reduced form of the mediator should not readily react with oxygen.

The ferrocenes represent a commonly used family of mediators (Figure 6.7(*a*)). Their reactions may be represented as follows:

$$
\begin{array}{ccccc}
\beta\text{-D-glucose} & & \text{FAD} & & \text{ferrocene} \quad +\,2\text{H}^+ \\
& & & & 2\text{Fe} \\
& \text{glucose oxidase} & & & \\
\text{D-glucono-1,5-lactone} & & \text{FADH}_2 & & 2\text{Fe}^+ \\
& & & & \text{ferricinium ion}
\end{array} \qquad [\mathbf{6.17}]
$$

$$2e^-\;\text{electrode}$$

Electrodes have now been developed which can remove the electrons directly from the reduced enzymes, without the necessity for such mediators. They utilise a coating of electrically conducting organic salts, such as

N-methylphenazinium cation (NMP$^+$, Figure 6.7(*b*)) with tetracyanoquino-dimethane radical anion (TCNQ$^-$, Figure 6.7(*c*)). Many flavoenzymes are strongly adsorbed by such organic conductors due to the formation of salt links, utilising the alternative positive and negative charges, within their hydrophobic environment. Such enzyme electrodes can be prepared by simply dipping the electrode into a solution of the enzyme and they may remain stable for several months. These electrodes can also be used for reactions involving NAD(P)$^+$-dependent dehydrogeneses, as they also allow the electrochemical oxidation of the reduced forms of these coenzymes. The three types of amperometric biosensor-utilising product, mediator or organic conductors represent the three generations in biosensor development (Figure 6.8). The reduction in oxidation potential, found when mediators are used, greatly reduces the problem of interference by extraneous material.

The current (i) produced by such amperometric biosensors is related to the rate of reaction (v_A) by the expression:

$$i = nFAv_A \tag{6.6}$$

where n represents the number of electrons transferred, A is the electrode area, and F is the Faraday. Usually the rate of reaction is made diffusionally controlled (see equation (3.27), p. 106) by use of external membranes. Under these circumstances the electric current produced is proportional to the analyte concentration and is independent both of the enzyme and electrochemical kinetics.

Optical biosensors

There are two main areas of development in optical biosensors. These involve determining changes in light absorption between the reactants and products of a reaction, or measuring the light output by a luminescent process. The former usually involve the widely established, if rather low technology, use of colorimetric test strips. These are disposable single-use cellulose pads impregnated with enzyme and reagents. The most common use of this technology is for a whole-blood monitoring in diabetes control. In this case, the strips include glucose oxidase, horseradish peroxidase (EC 1.11.1.7) and a chromogen (e.g. *o*-toluidine or 3,3′,5,5′-tetramethylbenzidine). The hydrogen peroxide, produced by the aerobic oxidation of glucose (see reaction scheme [1.1]), oxidising the weakly coloured chromogen to a highly coloured dye:

$$\text{chromogen (2H)} + H_2O_2 \xrightarrow{\text{peroxidase}} \text{dye} + 2H_2O \tag{6.24}$$

The evaluation of the dyed strips is best achieved by the use of portable reflectance meters, although direct visual comparison with a coloured chart is

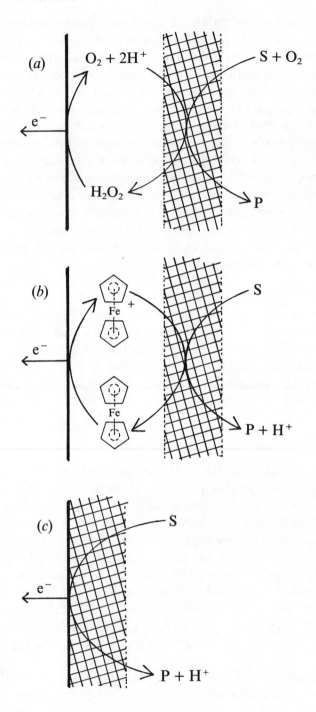

often used. A wide variety of test strips involving other enzymes are commercially available at the present time.

A most promising biosensor involving luminescence uses firefly luciferase (*Photinus*-luciferin 4-monooxygenase (ATP-hydrolysing), EC 1.13.12.7) to detect the presence of bacteria in food or clinical samples. Bacteria are specifically lysed and the ATP released (roughly proportional to the number of bacteria present) reacted with D-luciferin and oxygen in a reaction which produces yellow light in high quantum yield.

$$\text{ATP} + \text{D-luciferin} + O_2 \xrightarrow{\text{luciferase}} \text{oxyluciferin} + \text{AMP} +$$
$$\text{pyrophosphate} + CO_2 + \text{light (562 nm)} \quad [\textbf{6.25}]$$

The light produced may be detected photometrically by use of high-voltage, and expensive, photomultiplier tubes or low-voltage cheap photodiode systems. The sensitivity of the photomultiplier-containing systems is, at present, somewhat greater ($< 10^4$ cells ml^{-1}, $< 10^{-12}$M ATP) than the simpler photon detectors, which use photodiodes. Firefly luciferase is a very expensive enzyme, only obtainable from the tails of wild fireflies. Use of immobilised luciferase greatly reduces the cost of these analyses.

Figure 6.8. Amperometric biosensors for flavo-oxidase enzymes illustrating the three generations in the development of a biosensor. The biocatalyst is shown schematically by the cross-hatching. (*a*) First-generation electrode utilising the H_2O_2 produced by the reaction ($E^\circ = +0.68$ V). (*b*) Second-generation electrode utilising a mediator (ferrocene) to transfer the electrons, produced by the reaction, to the electrode ($E^\circ = +0.19$ V). (*c*) Third-generation electrode directly utilising the electrons produced by the reaction ($E^\circ = +0.10$ V). All electrode potentials (E°) are relative to the Cl^-/AgCl,Ag electrode. The following reaction occurs at the enzyme in all three biosensors:

$$\text{substrate(2H)} + \text{FAD-oxidase} \rightarrow \text{product} + \text{FADH}_2\text{-oxidase} \quad [\textbf{6.18}]$$

This is followed by the processes:

(*a*)
Biocatalyst:
$$\text{FADH}_2\text{-oxidase} + O_2 \rightarrow \text{FAD-oxidase} + H_2O_2 \quad [\textbf{6.19}]$$

Electrode:
$$H_2O_2 \rightarrow O_2 + 2H^+ + 2e^- \quad [\textbf{6.20}]$$

(*b*)
Biocatalyst:
$$\text{FADH}_2\text{-oxidase} + 2 \text{ ferricinium}^+ \rightarrow \text{FAD-oxidase} + 2 \text{ ferrocene} + 2H^+$$
$$[\textbf{6.21}]$$

Electrode:
$$2 \text{ ferrocene} \rightarrow 2 \text{ ferricinium}^+ + 2e^- \quad [\textbf{6.22}]$$

(*c*)
Biocatalyst/electrode:
$$\text{FADH}_2\text{-oxidase} \rightarrow \text{FAD-oxidase} + 2H^+ + 2e^- \quad [\textbf{6.23}]$$

Piezo-electric biosensors

Piezo-electric crystals (e.g. quartz) vibrate under the influence of an electric field. The frequency of this oscillation (f) depends on their thickness and cut, each crystal having a characteristic resonant frequency. This resonant frequency changes as molecules adsorb or desorb from the surface of the crystal, obeying the relationship:

$$\Delta f = K f^2 \Delta m / A \qquad (6.7)$$

where Δf is the change in resonant frequency (Hz), Δm is the change in mass of adsorbed material (g), K is a constant for the particular crystal dependent on such factors as its density and cut, and A is the adsorbing surface area (cm^2). For any piezo-electric crystal, the change in frequency is proportional to the mass of absorbed material, up to about a 2% change. This frequency change is easily detected by relatively unsophisticated electronic circuits. A simple use of such a transducer is a formaldehyde biosensor, utilising a formaldehyde dehydrogenase coating immobilised to a quartz crystal and sensitive to gaseous formaldehyde. The major drawback of these devices is the interference from atmospheric humidity and the difficulty in using them for the determination of material in solution. They are, however, inexpensive, small and robust, and capable of giving a rapid response.

Immunosensors

Biosensors may be used in conjunction with *enzyme-linked immunosorbent assays* (ELISA). The principles behind the ELISA technique are shown in Figure 6.9. ELISA is used to detect and amplify an antigen–antibody reaction; the amount of enzyme-linked antigen bound to the immobilised antibody being determined by the relative concentration of the free and conjugated antigen and quantified by the rate of enzymic reaction. Enzymes with high turnover numbers are used in order to achieve rapid response. The sensitivity of such assays may be further enhanced by utilising enzyme-catalysed reactions which give intrinsically greater response, for instance those giving rise to highly coloured, fluorescent or bioluminescent products. Assay kits using this technique are now available for a vast range of analyses.

Recently ELISA techniques have been combined with biosensors, to form *immunosensors*, in order to increase their range, speed and sensitivity. A simple immunosensor configuration is shown in Figure 6.10 (a), where the biosensor merely replaces the traditional colorimetric detection system. However, more advanced immunosensors are being developed (Figure 6.10 (b)); they rely on the direct detection of antigen bound to the antibody-coated

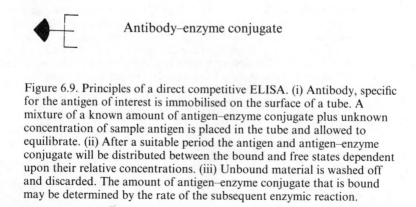

Figure 6.9. Principles of a direct competitive ELISA. (i) Antibody, specific for the antigen of interest is immobilised on the surface of a tube. A mixture of a known amount of antigen–enzyme conjugate plus unknown concentration of sample antigen is placed in the tube and allowed to equilibrate. (ii) After a suitable period the antigen and antigen–enzyme conjugate will be distributed between the bound and free states dependent upon their relative concentrations. (iii) Unbound material is washed off and discarded. The amount of antigen–enzyme conjugate that is bound may be determined by the rate of the subsequent enzymic reaction.

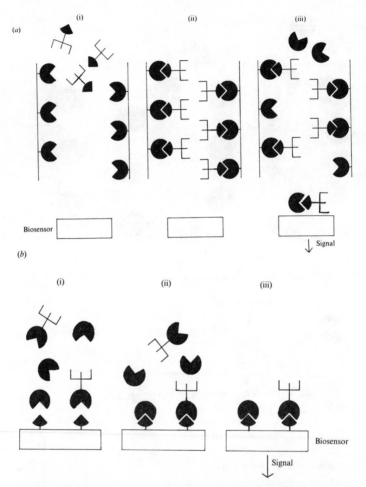

Figure 6.10. Principles of immunosensors. (*a*) (i) A tube is coated with (immobilised) antigen. An excess of specific antibody–enzyme conjugate is placed in the tube and allowed to bind. (*a*). (ii) After a suitable period any unbound material is washed off. (*a*) (iii) The analyte antigen solution is passed into the tube, binding and releasing some of the antibody–enzyme conjugate, depending upon the concentration of antigen. The amount of antibody–enzyme conjugate released is determined by the response from the biosensor. (*b*) (i) A transducer is coated with (immobilised) antibody, specific for the antigen of interest. The transducer is immersed in a solution containing a mixture of a known amount of antigen–enzyme conjugate plus unknown concentration of sample antigen. (*b*) (ii) After a suitable period the antigen and antigen–enzyme conjugate will be distributed between the bound and free states dependent upon their relative concentrations. (*b*) (iii) Unbound material is washed off and discarded. The amount of antigen–enzyme conjugate bound is determined directly from the transduced signal.

surface of the biosensor. Piezo-electric and FET-based biosensors are particularly suited to such applications.

Summary

(*a*) There is a vast potential market for biosensors which is only beginning to be exploited.

(*b*) Biosensors generally are easy to operate, analyse over a wide range of useful analyte concentrations and give reproducible results.

(*c*) The diffusional limitation of substrate(s) may be an asset to be encouraged in biosensor design due to the consequent reduction in the effects of analyte pH, temperature and inhibitors on biosensor response.

Bibliography

Albery, J., Haggett, B. & Snook, D. (1986). You know it makes sensors. *New Scientist*, 13 February, pp. 38–41.

Bergmeyer, H. U. (ed.) (1974). *Methods of enzymatic analysis*, 3rd edn. New York: Verlag Chemie, Academic Press.

Guilbault, G. G. & de Olivera Neto, G. (1985). Immobilised enzyme electrodes. In *Immobilised cells and enzymes: a practical approach*, ed. J. Woodward, pp. 55–74. Oxford: IRL Press Ltd.

Hall, E.A.H. (1986). The developing biosensor arena. *Enzyme and Microbial Technology*, **8**, 651–7.

Joachim, C. (1986). Biochips: dreams ... and realities. *International Industrial Biotechnology*, no. 79:7:12/1, pp. 211–19.

Kernevez, J. P., Konate, L. & Romette, J. L. (1983). Determination of substrate concentration by a computerized enzyme electrode. *Biotechnology and Bioengineering*, **25**, 845–55.

Kricka, L. J. & Thorpe, G. H. G. (1986). Immobilised enzymes in analysis. *Trends in Biotechnology*, **4**, 253–8.

Lowe, C. R. (1984). Biosensors. *Trends in Biotechnology*, **2**, 59–64.

North, J. R. (1985). Immunosensors: antibody-based biosensors. *Trends in Biotechnology*, **3**, 180–6.

Russell, L. J. & Rawson, K. M. (1986). The commercialisation of sensor technology in clinical chemistry: an outline of the potential difficulties. *Biosensors*, **2**, 301–18.

Scheller, F. W., Schubert, F., Renneberg, R. & Müller, H.-G. (1985). Biosensors: trends and commercialization. *Biosensors*, **1**, 135–60.

Turner, A. P. F. (1987). Biosensors: principles and potential. In *Chemical aspects of food enzymes*, ed. A. T. Andrews, pp. 259–70. London: Royal Society of Chemistry.

Turner, A. P. F., Karube, I. & Wilson, G. S. (eds.) (1987). *Biosensors: fundamentals and applications*. Oxford: Oxford University Press.

Vadgama, P. (1986). Urea pH electrodes: characterisation and optimisation for plasma measurements. *Analyst*, **111**, 875–8.

7 Recent advances in enzyme technology

Enzymic reactions in biphasic liquid systems

It would often be useful if enzyme-catalysed reactions could be performed in solvents other than water, as this is not the ideal medium for the majority of organic reactions. Many reactants (e.g. molecular oxygen, steroids and lipids) are more soluble in organic solvents than in water and some products may be quite labile in an aqueous environment. Accomplishing reactions with such substrates (e.g. the aerobic oxidation of oestrogens catalysed by fungal laccase) in non-aqueous media allows a much increased volumetric activity to be achieved. Microbial contamination, by contrast, is much less of a problem in such solvents, and the consequent absence of microbial proteases may lead to an apparent stabilisation in the biocatalyst.

Some polymerising reactions, for example the polymerisation of phenols catalysed by peroxidase, will produce a higher molecular weight product when carried out in a solution more able to dissolve the products (i.e. oligomers) initially formed. Under normal physiological conditions, hydrolytic enzymes catalyse the degradation of polymers; i.e. hydrolases are transferases normally transferring a moiety to the acceptor, water. Water is normally present in vast molar excess over other potential acceptor molecules, so no reaction other than hydrolysis occurs. Also, the normal 'concentration' of water (about 55.5 M) is much greater than its typical K_m (about 50 mM) and the rate of hydrolysis will not be affected as the reaction proceeds. By greatly reducing the water activity in these systems they can be used to transfer to other acceptors. Examples of this can be found in the transesterification reactions of esterases and lipases, described more fully later, and the (undesirable) formation of isomaltose from glucose catalysed by glucoamylase.

Restriction of the enzyme to the aqueous phase effectively immobilises the enzyme and allows its straightforward separation, using phase separators developed for the established chemical process industry, from product-containing organic phase. The main asset of these systems, however, is their ability to shift the thermodynamic equilibria of the reactions. As was pointed

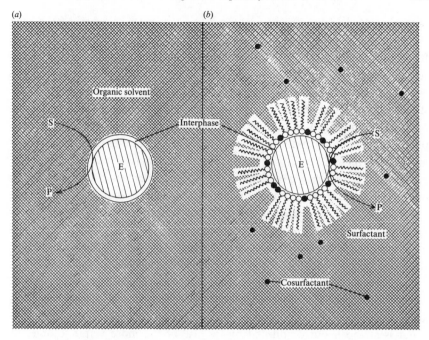

Figure 7.1. Diagram showing two configurations for an enzyme within an organic solvent. (*a*) Almost-anhydrous enzyme suspended in the organic solvent. The enzyme (E) is surrounded by a thin interphase consisting of water or water plus immobilisation support. (*b*) Enzyme dissolved in a reversed micellar medium. The micelles are formed by the surfactant molecules with assistance from the co-surfactant (if present). The surfactant (e.g. cetyltrimethylammonium bromide (CTAB), bis(2-ethylhexyl) sodium sulphosuccinate (AOT), phosphatidylcholine, tetraethyleneglycoldodecylether) is found only at the interphase boundary, whereas the water-immiscible co-surfactant (e.g. butanol, hexanol, octanol), added to vary the properties of this interphase, is generally less polar and more soluble in the organic continuous phase. Both preparations ((*a*) and (*b*)) give optically transparent solutions.

out in Chapter 1, enzymes do not change the equilibrium constants (K_{eq}) of reactions. Although changes in the physical conditions (i.e. temperature, pH and pressure) do affect the K_{eq} of a reaction, usually this effect is relatively slight over the physical range allowed by stability of the biocatalysts. Use of a biphasic aqueous-organic system, however, may result in substantial changes in the practically useful 'apparent' K_{eq}. The use of enzymes within organic solvents normally results in a two-phase system as all water-soluble enzymes possess a significant amount of strongly bound water, even when in an apparently dry state. This is shown schematically in Figure 7.1.

The stabilisation of enzymes in biphasic aqueous–organic systems

It should become clear from the later discussion that there may be a substantial advantage to be gained from the use of biphasic systems in many enzyme-catalysed reactions. One major factor must first be addressed, the stability of the enzyme in these systems. A distinction should be drawn between the more water-soluble hydrophilic enzymes and the more hydrophobic enzymes often associated with lipid and membranes (e.g. lipases). The active integrity and stability of hydrophilic enzymes appear to depend on the presence of a thin layer of water, just a few molecules thick, within the microenvironment. This amount of water is miniscule (between 50 and 500 molecules of water for each enzyme molecule) and the enzyme may effectively be operating in an almost anhydrous state. Some hydrophobic lipases retain activity even if fewer molecules of water remain; presumably just sufficient to stabilise the conformation of the active site. The pH of such minute pools of water, containing no free hydrogen ions, is impossible to measure, or control, directly. However, it appears that the enzyme 'remembers' the pH of its last aqueous solution and functions as though at that pH. If the enzyme-bound water is stripped out or diluted by the use of the more water-soluble, or miscible, organic solvents then the enzyme is usually inactivated. However, under conditions where this does not occur, the limited amount of water available, and the associated reduction in the water activity, considerably reduce the rate of thermoinactivation. This has a stabilising effect on most enzymes; porcine pancreatic lipase, for example, has a half-life of more than 12 h at 100°C in 0.02% water in tributyrin, whereas this drops to 12 min at a 0.8% water content and inactivation is almost instantaneous in 100% water. Additionally, the freezing point of the water is reduced, which allows the use of particularly heat-labile enzymes at very low temperatures. The lowering of the water activity tends to produce a more rigid enzyme molecule, which may affect both the K_m and V_{max} of the enzyme. In extreme cases, this may result in a change in the catalytic properties. Porcine pancreatic lipase demonstrates this effect. When used in biphasic systems of low water activity, it no longer catalyses transesterification reactions involving the bulky tertiary alcohols.

The most important factor in the balance between stabilisation and inactivation, due to organic phase, is the solvent polarity. Solvents of lower polarity (i.e. greater hydrophobicity) are less able to disrupt the structure of the necessary tightly bound water molecules. The best measure of polarity is the logarithm of the partition coefficient ($\log P$) of the organic liquid between *n*-octanol and water; the higher the $\log P$, the more non-polar (hydrophobic) is the solvent (Table 7.1):

Table 7.1   *Log* P *values of the more commonly used organic solvents*

Solvent	Log *P*	Solvent	Log *P*
Butanone	0.3	1,1,1-Trichloroethane	2.8
Ethyl acetate	0.7	Carbon tetrachloride	2.8
Butanol	0.8	Dibutyl ether	2.9
Diethyl ether	0.8	Cyclohexane	3.1
Methylene chloride	1.4	Hexane	3.5
Butyl acetate	1.7	Petroleum ether (60–80)	3.5
Diisopropyl ether	2.0	Petroleum ether (80–100)	3.8
Benzene	2.0	Dipentyl ether	3.9
Chloroform	2.2	Heptane	4.0
Tetrachloroethylene	2.3	Petroleum ether (100–120)	4.3
Toluene	2.7	Hexadecane	8.7

Log P values increase by about 0.52 for every methylene group (-CH$_2$-) added in a homologous series. Thus, the log P of hexanol is that of butanol (0.8) plus 2 × 0.52 (i.e. approximately 1.8).

$$\log P = \log\left(\frac{[\text{Material}_{\text{octanol}}]}{[\text{Material}_{\text{water}}]}\right) \tag{7.1}$$

There appears to be a clear correlation between the activity of biocatalysts in two-phase systems and the log P (Figure 7.2). The S-shape of this relationship suggests that enzymes are generally inactivated by solvents with log $P < 2$ but are little affected by solvents with log $P > 4$. There is some variation in the effects between different enzymes and different solvents which makes activity prediction in the log P range 2–4 rather imprecise. This range includes some of the most utilised organic phases (e.g. chloroform), which may be suitable for some applications but cause harmful inactivation in others. The solubility of the reactant(s) and product(s) may considerably reduce the range of log P that is available for a particular application; many basically non-polar molecules possess some polar structural regions which cause their lack of solubility in strongly hydrophobic solvents. The choice of organic phase will also depend on additional factors such as cost, ease of recovery, fire and fume hazards, and specific inhibitory effects. The S-shaped curve can be shifted to the left by immobilising the enzyme within a highly hydrophilic support (Figure 7.2). A simple way of achieving this is to impregnate a beaded hydrophilic polymer (e.g. Sephadex, agarose) with the enzyme, followed by suspension of the wet beads directly in the organic phase. These shifts are very important as they greatly increase the choice of suitable organic phase. Such impregnated beads have the additional advantages that: (1) they protect the contained enzyme from liquid–liquid interfacial

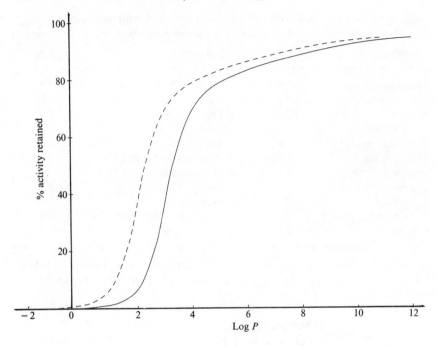

Figure 7.2. Diagram showing the dependence of the activity of
immobilised enzymes, in biphasic systems, on the log*P* of the organic
phase. ——— Free enzyme; ————— enzyme immobilised within a
strongly hydrophilic support.

denaturation at higher rates of stirring; (2) they enable more facile recovery
of the biocatalyst; (3) they may be used with low molecular weight hydrophilic coenzymes (e.g. $NAD(P)^+$) with the assistance of a coenzyme-regenerating process, the coenzyme being effectively immobilised within the aqueous
pools; and (4) they allow the efficient and continuous use of biphasic PBRs,
so long as the moving organic phase remains saturated with water.

Biphasic systems may be further stabilised by the use of deuterated water
(^2H_2O). This reduces the rate of thermal inactivation, although it does cause
an increase in the pK_a values of ionising groups by about 0.4, with the
associated changes in the pH–activity and pH–stability relationships. The
higher cost of the deuterated water is offset, to a certain extent, by the small
amount necessary in these systems and the ease with which it may be
recovered at the end of a catalytic process.

Even where the organic solvent has very low $\log P$ and is miscible, the
effect of the expected loss in enzymic activity may be offset by changes in the
equilibrium constant. Thus it has been proposed that glucose isomerase be

used in aqueous ethanol to produce high-fructose corn syrup. The equilibrium fraction of fructose can be raised by about 10% to 55% (w/w) at 30 °C in 80% (v/v) ethanol. This is economically valid even though the enzymic activity drops by about 10% compared to that in the absence of ethanol.

Equilibria in biphasic aqueous–organic systems

Much of the theory outlining the effect of biphasic systems on the apparent equilibria was outlined by Karel Martinek and his coworkers at Lomonosov Moscow State University. Consider the simple reaction scheme, involving the equilibrium between reactant (A) and product (B), K_W being the equilibrium constant (K_{eq}) in water:

$$A \underset{}{\overset{K_W}{\rightleftharpoons}} B \qquad\qquad [7.1]$$

If this reaction is carried out in a biphasic system consisting of a mixture of an aqueous and an organic solvent, both A and B will be partitioned between the two solvents (with partition coefficients, P_A and P_B) and a separate equilibrium (K_{org}) will be established between A and B in the organic phase:

$$
\begin{array}{ll}
\text{aqueous phase} & A \overset{K_W}{\rightleftharpoons} B \\
& P_A \Big\Updownarrow \quad \Big\Updownarrow P_B \\
\text{organic phase} & A \underset{K_{org}}{\rightleftharpoons} B
\end{array}
\qquad\qquad [7.2]
$$

where:

$$K_W = [B_W]/[A_W] \qquad\qquad (7.2)$$

$$K_{org} = [B_{org}]/[A_{org}] \qquad\qquad (7.3)$$

$$P_A = [A_{org}]/[A_W] \qquad\qquad (7.4)$$

$$P_B = [B_{org}]/[B_W] \qquad\qquad (7.5)$$

Because of the cyclic nature of these equilibria, only three of these parameters are independent 'variables' (i.e. K_{org} depends on K_w, P_A and P_B). It should be noticed from the following discussion that K_{org} plays no further part, although similar derivations could be made involving K_{org}, P_A and P_B rather than K_w, P_A and P_B. The apparent equilibrium constant ($K_{biphasic}$) of this system may be defined as

$$K_{biphasic} = [B_t]/[A_t] \qquad\qquad (7.6)$$

where the subscripts t, org and W refer to the total solution, the organic and water phases, respectively. If V represents the volumes involved, it follows that:

$$[A_t]V_t = [A_W]V_W + [A_{org}]V_{org} \tag{7.7}$$

and:

$$[B_t]V_t = [B_W]V_W + [B_{org}]V_{org} \tag{7.8}$$

where:

$$V_t = V_W + V_{org} \tag{7.9}$$

Substituting from equations (7.7) and (7.8) into equation (7.6):

$$K_{biphasic} = \frac{[B_W]V_W + [B_{org}]V_{org}}{[A_W]V_W + [A_{org}]V_{org}} \tag{7.10}$$

Dividing both the numerator and denominator by $[A_W]V_W$, this becomes:

$$K_{biphasic} = \frac{([B_W]/[A_W]) + [B_{org}]/[B_W])([B_W]/[A_W])(V_{org}/V_W)}{1 + ([A_{org}]/[A_W])(V_{org}/V_W)} \tag{7.11}$$

Substituting the partition coefficients from equations (7.4) and (7.5) gives the simplified relationship:

$$K_{biphasic} = K_W \frac{(1 + \alpha P_B)}{(1 + \alpha P_A)} \tag{7.12}$$

where:

$$\alpha = V_{org}/V_W \tag{7.13}$$

i.e. α is the ratio of the volumes of the organic and water phases.

The apparent equilibrium constant ($K_{biphasic}$) varies with the relative volumes of the aqueous and organic phases, increasing when the substrate is

Figure 7.3. The variation in the apparent equilibrium constants of a one-substrate one-product reaction ($A \rightleftharpoons B$) with the relative component composition of a biphasic aqueous–organic system. (a) the effect of the ratio of the partition coefficients, P_B/P_A. The curves have been generated using equation (7.12) and the following values for the partition coefficients, from the top downwards: $P_A = 0.2$, $P_B = 20$; $P_A = 0.2$, $P_B = 2$; $P_A = 0.2$, $P_B = 0.6$; $P_B = 0.6$, $P_B = 0.2$; $P_A = 2$, $P_B = 0.2$; $P_A = 20$, $P_B = 0.2$ (b) the effect of the polarity of A and B, keeping the ratio of the partition coefficients constant ($P_B/P_A = 100$). The following values for the partition coefficients have been used, from the top downwards: $P_B = 1000$, $P_B = 100$, $P_B = 10$, $P_B = 1$, $P_B = 0.1$, $P_A = 0.01$, $P_A = 0.001$, $P_B = 0.0001$.

(a)

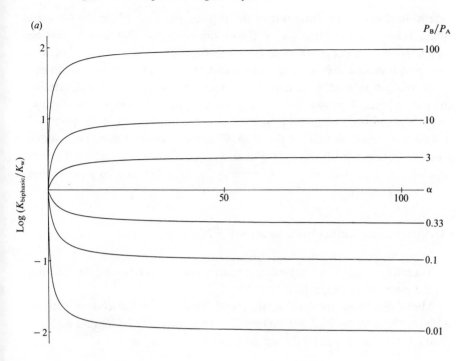

(b)

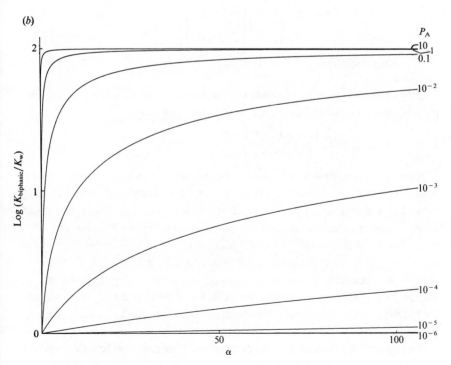

partitioned more efficiently out of the organic phase and into the aqueous phase relative to the behaviour of the product (i.e. $P_A < P_B$), and decreasing when the reverse occurs (Figure 7.3(a)). If there are sufficient differences in the partition coefficients and both substrate and product are relatively non-polar, then substantial shifts in the equilibria occur even at high relative water contents. A substantial increase in the apparent equilibrium constant may be achieved when both the substrate and product have partition coefficients less than unity, if the ratio of the partition coefficients is suitable (Figure 7.3(b)). Clearly if α is very small $K_{biphasic}$ tends to K_W, the equilibrium constant in aqueous solution. However, if α is sufficiently large, equation (7.12) simplifies to:

$$K_{biphasic} = K_W P_B / P_A \qquad (7.14)$$

Therefore, substituting from equations (7.2)–(7.5):

$$K_{biphasic} = [B_{org}]/[A_{org}] = K_{org} \qquad (7.15)$$

Therefore, $K_{biphasic}$ tends to K_{org}, the equilibrium constant of the reaction in the pure organic liquid.

Many reactions are bimolecular, and these are influenced in a more complex way by the biphasic system. The processes may be represented in a manner similar to that used for monomolecular reactions:

$$
\begin{array}{llll}
\text{aqueous phase} & A + B \overset{K_W}{\rightleftharpoons} & C + D & \\
& P_A \updownarrow \quad \updownarrow P_B \quad P_C \updownarrow \quad \updownarrow P_D & & [7.3] \\
\text{organic phase} & A + B \underset{K_{org}}{\rightleftharpoons} & C + D &
\end{array}
$$

Using reasoning similar to that outlined above, the following expression may be obtained for the apparent equilibrium constant:

$$K_{biphasic} = K_w \frac{(1 + \alpha P_C)(1 + \alpha P_D)}{(1 + \alpha P_A)(1 + \alpha P_B)} \qquad (7.16)$$

As in the case of monomolecular reactions, when the relative water content is very low (i.e. α is high) the apparent equilibrium constant tends to K_{org}. However, the equation (7.16) is quadratic in terms of α and at intermediate values may produce a maximum or minimum value for the apparent equilibrium constant that is several orders of magnitude different from either K_w or K_{org} (i.e. the apparent equilibrium constant may be greater than the equilibrium constant of the same reaction in either of the pure phases, see Figure 7.4). Under some circumstances this may enable the reaction productivity in the biphasic system to be much greater than that attainable in either pure phase.

A special and relevant case of the bimolecular reaction scheme is where

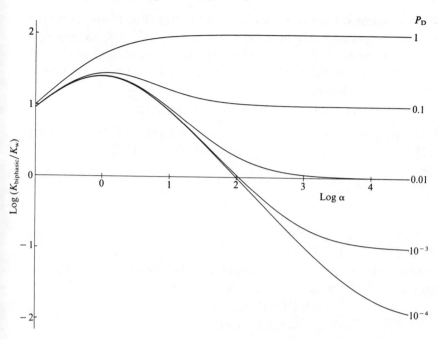

Figure 7.4. The variation in the apparent equilibrium constants of a two-substrate two-product reaction (A + B ⇌ C + D) with the relative component composition of a biphasic aqueous–organic system. In all cases the partition coefficients are $P_A = 1$, $P_B = 1$ and $P_C = 100$. The following values for the partition coefficient (P_D) have been used, from the top downwards; 1, 0.1, 0.01, 0.001 and 0.0001.

one of the reactants is water (e.g. use of the hydrolases). These reactions may be written in a way corresponding to reaction scheme [**7.3**]:

$$\text{aqueous phase} \quad A + B \underset{\displaystyle K_W}{\overset{\displaystyle \text{hydrolase}}{\rightleftharpoons}} C + H_2O$$

$$P_A \Updownarrow \quad \Updownarrow P_B \quad P_C \Updownarrow \quad \Updownarrow P_W \qquad [\textbf{7.4}]$$

$$\text{organic phase} \quad A + B \underset{\displaystyle K_{org}}{\rightleftharpoons} C + H_2O$$

The normal direction for the reaction in aqueous solution is the hydrolytic process from right to left. However, the apparent equilibrium constant may be shifted in biphasic solutions, as outlined above, and allow the hydrolase to act in synthetic (i.e. left to right in reaction scheme [**7.4**]), rather than hydrolytic, manner. Such reactions significantly extend the processes available to the enzyme technologist, as synthetic reactions are generally much more difficult to achieve by classical organic chemistry than are hydrolytic reactions; this is true particularly when a regiospecificity is required, as then there is a choice of reactive groups in the substrates. Two factors affect the yield of the product (C) in such reactions: (1) the shift in apparent equilibrium constant ($K_{biphasic}$); and (2) the total concentration of water ($[(H_2O)_t]$), as one of the participating reactants. $[(H_2O)_t]$ may be obtained from the material balance:

$$[(H_2O)_t](V_w + V_{org}) = [(H_2O)_w]V_w + [(H_2O)_{org}]V_{org} \qquad (7.17)$$

where $[(H_2O)_w]$ is the molar concentration of pure water ($= 1000/18 = 55.5$ M). Substituting P_W for $[(H_2O)_{org}]/[(H_2O)_w]$:

$$[(H_2O)_t] = \frac{55.5(1 + \alpha P_W)}{1 + \alpha} \qquad (7.18)$$

but:

$$K_{biphasic} = \frac{[A_t][B_t]}{[C_t][(H_2O)_t]} \qquad (7.19)$$

Substituting for $K_{biphasic}$ from equation (7.16) (with P_D set to equal P_w) and rearranging, the terms in $(1 + \alpha P_w)$ cancel to give:

$$[C_t] = K_w \frac{(1 + \alpha P_C)(1 + \alpha)[A_t][B_t]}{55.5(1 + \alpha P_A)(1 + \alpha P_B)} \qquad (7.20]$$

This represents a fairly complex relationship, but a few generalisations may be made. If the product (C) is less polar than either reactant (A or B), then the yield generally increases with the relative concentration of the organic phase (α) to reach a plateau. If the reverse occurs, then there will be a plateau at low yield and high α in the α-yield diagram. Both minima and maxima may occur, depending on the parameters. Figure 7.5 shows the variation of the apparent equilibrium constant and the yield for the enzymic synthesis of an ester from its acid and alcohol in a biphasic system. From this it can be seen that the use of a biphasic system can increase the yield of such reactions from zero to nearly 100%. In order to achieve high conversions, the water produced must be removed from the two-phase system. If it is not removed, there is a continual drop in α, which reduces the extent of the favourable equilibrium. There are no simple ways in which this water may be removed but reactors which allow the constant addition of fresh catalyst and organic phase reduce the size of the problem.

Use of biphasic solvent systems also affect the ionisation of acid and basic groups. Consider the ionisation of an acid HA, where only the unionised form is soluble in the organic phase and the ionic species are only soluble in the aqueous phase:

$$\text{aqueous phase} \qquad HA \underset{K_{a,w}}{\rightleftharpoons} H^+ + A^-$$

$$P_{HA} \updownarrow$$

$$\text{organic phase} \qquad HA \qquad\qquad\qquad [7.5]$$

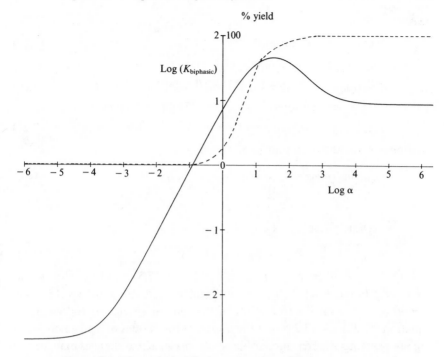

Figure 7.5. The variation in the apparent equilibrium constant and percentage yield of a reverse hydrolytic reaction (A + B$\rightleftharpoons$C + H$_2$O) with the relative component composition of a biphasic aqueous–organic system. The reaction portrayed is the synthesis of *N*-benzoyl-L-phenylalanine ethyl ester, from *N*-benzoyl-L-phenylalanine and ethyl alcohol, catalysed by α-chymotrypsin in a biphasic chloroform–water system (pH 7). ———— Apparent equilibrium constant; ———— % yield (semi-logarithmic plot). The curves were calculated assuming $K_w = 0.002$, $P_A = 0.11$ *N*-benzoyl-L-phenylalanine at pH 7, $P_B = 0.01$ (ethyl alcohol), $P_C = 4100$ (*N*-benzoyl-L-phenylalanine ethyl ester), and an equimolar mixture of reactants.

where:

$$K_{a,w} = [H^+][A_{\bar{w}}]/[HA_w] \tag{7.21}$$

therefore:

$$pK_{a,w} = pH + \log([HA_w]/[A_{\bar{w}}]) \tag{7.22}$$

The hydrogen ion concentration, as determined, is characteristic of the aqueous phase. Therefore:

$$pK_{a,biphasic} = pH + \log([HA_t]/[A_t^-]) \tag{7.23}$$

A consideration of material mass balances for A$^-$ and HA gives:

$$[A_t^-]V_t = [A_{\bar{w}}]V_w \tag{7.24}$$

and:

$$[HA_t]V_t = [HA_W]V_W + [HA_{org}]V_{org} \tag{7.25}$$

Substituting for pH from equation (7.22) into equation (7.23) gives:

$$pK_{a,biphasic} = pK_{a,W} + \log\{([HA_t]/[A_t^-])([A_W^-]/[HA_W])\} \tag{7.26}$$

Substituting from equations (7.24) and (7.25) this simplifies to give:

$$pK_{a,biphasic} = pK_{a,W} + \log(1 + \alpha P_{HA}) \tag{7.27}$$

Similarly, for the protonation of a base:

$$\text{aqueous phase} \qquad \text{base} + H^+ \underset{}{\overset{K_a}{\rightleftharpoons}} \text{baseH}^+$$

$$P_B \Big\updownarrow \qquad\qquad\qquad\qquad\qquad \tag{7.6}$$

$$\text{organic phase} \qquad \text{base}$$

$$pK_{a,biphasic} = pK_{a,W} - \log(1 + \alpha P_{base}) \tag{7.28}$$

A qualitative assessment of reaction schemes [7.5] and [7.6] shows that increasing the partition coefficient for the uncharged acid or base will pull the reaction away from the formation of the ionised species. It follows from equations (7.27) and (7.28) that the apparent pK_a values of acids increase in biphasic systems whereas those of bases decrease. These changes may well be several pH units, dependent on the partition coefficients and the relative fraction of organic solvent present. The shifts in pK_a are in addition to any increase in the pK_a due to the lower dielectric constant of the aqueous phase, as outlined in Chapter 1. This is likely to be particularly relevant under almost anhydrous conditions (i.e. where α is high) which tend to 'freeze' the hydrogen ions, lowering their activity. The utility of the shift in the pK_a can be shown using ester synthesis catalysed by α-chymotrypsin as an example.

$$\text{aqueous phase} \qquad RCOO^- + H^+$$

$$K_{a,acid} \Big\updownarrow \qquad\qquad\qquad\qquad\qquad\qquad\qquad [7.7]$$

$$\text{aqueous phase} \qquad RCOOH + R'OH \underset{}{\overset{K_{non\text{-}ionic,W}}{\rightleftharpoons}} RCOOR' + H_2O$$

$$P_{acid} \Big\updownarrow \qquad\quad \Big\updownarrow P_{alcohol} \qquad P_{ester} \Big\updownarrow \qquad \Big\updownarrow P_W$$

$$\text{organic phase} \qquad RCOOH + R'OH \underset{K_{non\text{-}ionic,org}}{\rightleftharpoons} RCOOR' + H_2O$$

where RCOOH and RCOO$^-$ are the unionised and ionised forms of N-benzoyl-L-phenylalanine, RCOOR$'$ is N-benzoyl-L-phenylalanine ethyl ester and R$'$OH is ethyl alcohol.

Although the non-ionic reaction in aqueous solution is evenly balanced ($K_{non\text{-}ionic,W} = 7$), the overall reaction proceeds towards the left, in water at neutral pH, due to the pull exerted by the acid's ionisation ($pK_a = 3.4$). In

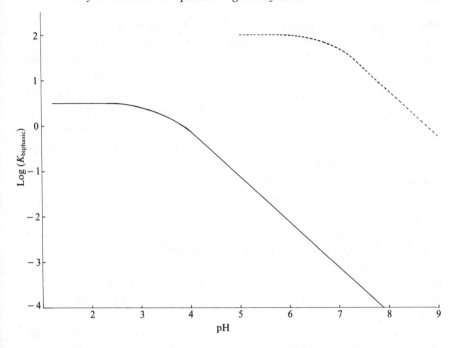

Figure 7.6. pH dependence of the apparent equilibrium constant for the synthesis of *N*-benzoyl-L-phenylalanine ethyl ester, as given by the reaction scheme [**7.7**]. ———— Reaction in aqueous solution only (i.e. $\alpha = 0$); –––––– reaction in a biphasic chloroform–water system ($\alpha = 20$). The shift upwards is due to the more significant partition of the ester into the organic phase relative to the reactants, whereas the shift in the pH dependence to higher pH is due to the change in the apparent pK_a of the acid.

biphasic chloroform–water solution ($\alpha = 20$, $P_{HA} = 100$), the K_a shifts by three orders of magnitude to give a pK_a close to 7. This, together with a shift in the equilibrium constant due to the partition effects outlined earlier, produces an overall shift in the equilibrium constant of about five orders of magnitude at the optimal pH (7.5) of the enzyme (Figure 7.6). Use of the biphasic system allows reactions to be performed at a pH that is both thermodynamically favourable to the reaction direction required and at the optimum for the enzyme's catalytic efficiency.

Enzyme kinetics in biphasic aqueous–organic systems

Although the equilibrium position of reactions may be shifted in biphasic systems, this is of no practical consequence unless the reaction actually occurs. The enzyme must be able to convert the reactants to products. For this to proceed, even in the presence of fully active enzymes, there must be some reactant present in the aqueous microenvironment surrounding the enzyme. Reactants may have fairly low aqueous solubilities under the normal conditions associated with such biphasic processes. The aqueous phase remains effectively unstirred, even when the organic phase is well stirred, during the reaction so that the thickness of the 'unstirred layer' (δ) is equal to the thickness of the aqueous layer surrounding the enzyme. Where δ is large, the rate of reaction is likely to be controlled by the passage of reactants through this aqueous layer, as the concentration gradients will be very shallow due to the low solubilities of the reactants in water. δ is often very thin, however, especially where the enzyme is freely soluble, causing the reactions to be effectively diffusionally controlled by the passage of the reactants from the more concentrated organic phase through the interphase boundary to very low concentrations within the microenviron-ment of the enzyme. Clearly, the thickness of the aqueous layer around an immobilised enzyme will be of great importance in determining the degree of diffusional resistance involved, and hence the lowering of the reaction rate. As with all immobilised enzymes, the volumetric surface area (A/V) is an important parameter governing the overall flux of the substrate to the biocatalytic surface, the effectiveness of the enzyme and the productivity. Whereas in reactions involving a single liquid phase this surface area is the area presented by the exterior of the particles to the bulk of the liquid, in most practical biphasic liquid systems it is the interfacial area between the two liquid phases that is relevant. Under normal operating conditions this interfacial area will depend upon the volume of the aqueous droplet surrounding and enclosing the enzymic biocatalyst. As the relative volume of the organic phase increases the volumetric surface area will decrease, so reducing the maximum volumetric productivity. This will be of particular importance where α needs to be very high in order to shift the reaction equilibrium. Clearly this will reduce the reaction rate due to the reduction in the amount of enzyme that can be present, due to its associated water, and the associated reduction in the interfacial area. Where the catalysed reaction generates water, this should not generally be allowed to accumulate as it not only has a detrimental effect on the equilibrium constant but will also slow down the rate of reaction.

There are some rules which allow the optimisation of the two-phase system for enzyme-catalysed reactions. They depend upon the concept of log P being extended to cover the substrates, products and the interphase (see Figure 7.1).

(1) The difference between the log P values of the substrate(s) and the interphase should be as small as possible, whereas that of the

substrate(s) should be much lower than that of the organic phase. These conditions encourage high concentrations of substrate(s) within the interphase and, hence, the transfer of the substrate from the organic phase into the aqueous phase.

(2) The difference between the $\log P$ values of the product(s) and the organic phase should be as small as possible, whereas that of the product(s) should be much greater than that of the interphase. These conditions encourage the transfer of the product from the aqueous phase through the interphase into the organic phase, after reaction.

The $\log P$ of the interphase can be varied independently of the organic phase by suitable choice of surfactants and co-surfactants. The precise structure of this interphase may be much more complex than that outlined in Figure 7.1. In particular, when there is excess water and surfactant present, the surfactant may form multiple concentric membranes, consisting of surfactant bilayers, around the aqueous core. This presents another reason for the removal of the water produced by use of hydrolases (used as synthetases) in biphasic systems. Control over the amount of water surrounding the enzyme is often possible by means of molecular sieves (e.g. potassium aluminium silicate), which absorb water. Removal of water dissolved in the organic phase can be achieved by their use and this leads to the depletion of water from the aqueous biocatalytic pools by its partition. If the organic phase has a high boiling point, the removal of water may be achieved by vacuum distillation, e.g. in the esterification of fatty acids and fatty alcohols catalysed by lipase and used in wax manufacture.

Use of aqueous two-phase systems

Aqueous two-phase systems (see Chapter 2) offer the opportunity to shift reaction equilibria towards product formation by ensuring that enzyme and substrate partition into one phase whilst the product enters, and may be removed from, the other. The theory developed (earlier) for aqueous–organic biphasic systems is equally applicable to these systems. An example of such an extractive bioconversion is a method for the conversion of starch to glucose using bacterial α-amylase and glucoamylase. Starch substrate partitions almost entirely into the lower, more hydrophilic, dextran-rich phase of a system comprising 3% polyethylene glycol (PEG 20000) and 5% crude unfractionated dextran. The enzymes also partition largely to the bottom phase but the glucose, produced by the hydrolysis, distributes itself more evenly between the phases. A small proportion of the glucoamylase enters the upper phase and will convert any oligosaccharide entering that phase to glucose. Concentrations of up to 140 g l^{-1} glucose may be reached in the upper phase.

Such a system offers some of the advantages of an immobilised-enzyme process without some of the disadvantages. The enzymes are largely retained and are stabilised by the presence of the polymers, yet catalysis is in homogeneous solution (within the phase) so no diffusion limitations to mass transfer exist. Drawbacks include the need to separate the product from the upper phase polymers and gradual loss of enzymes which enter the upper phase. Enzyme loss could be reduced, without introducing diffusion limitations, by linking them to hydrophilic polymers so as to form soluble complexes.

Practical examples of the use of enzymes 'in reverse'

It has long been known that if proteases are supplied with high concentrations of soluble proteins, peptides or amino acids, polymers (*plasteins*) are produced with apparently random, if rather hydrophobic, structures. This reaction has been used, for example, to produce bland-tasting, colourless plasteins from brightly coloured, unpleasant tasting, algal biomass and for the introduction of extra methionine into low quality soy protein. However in general, the non-specific use of proteases in the synthesis of new structures has not found commercial use. However, proteases have come into use as alternatives to chemical methods for the synthesis of peptides of known and predetermined structure because their specificity allows reactions to proceed stereospecifically and without costly protection of side-chains.

A method for the synthesis of the high intensity sweetener aspartame exemplifies the power of proteases (in this case, thermolysin). Aspartame is the dipeptide of L-aspartic acid with the methyl ester of L-phenylalanine (α-L-aspartyl-L-phenylalanyl-O-methyl ester). The chemical synthesis of aspartame requires protection of both the β-carboxyl group and the α-amino group of the L-aspartic acid. Even then, it produces aspartame in low yield and at high cost. If the β-carboxyl group is not protected, a cost saving is achieved but about 30% of the β-isomer is formed and has, subsequently, to be removed. When thermolysin is used to catalyse aspartame production the regiospecificity of the enzyme eliminates the need to protect this β-carboxyl group but the α-amino group must still be protected (usually by means of reaction with benzyl chloroformate to form the benzyloxycarbonyl (BOC) derivative, i.e. BOC-L-aspartic acid) to prevent the synthesis of poly (L-aspartic acid). More economical racemic amino acids can also be used, as only the desired isomer of aspartame will be formed.

If stoichiometric quantities of BOC-L-aspartic acid and L-phenylalanine methyl ester are reacted in the presence of thermolysin, an equilibrium

Table 7.2   *The specificity of some specific industrial proteases, involving acyl intermediates*

Enzyme	Preferred cleavage sites[a] (N-terminal→C-terminal)
Bromelain	—Lys$\overset{\downarrow}{-}$ Z; —Arg$\overset{\downarrow}{-}$ Z; —Phe$\overset{\downarrow}{-}$ Z; —Tyr$\overset{\downarrow}{-}$ Z
Chymotrypsin	—Trp$\overset{\downarrow}{-}$ Z; —Tyr$\overset{\downarrow}{-}$ Z; —Phe$\overset{\downarrow}{-}$ Z; —Leu$\overset{\downarrow}{-}$ Z
Papain	—Phe—AA$\overset{\downarrow}{-}$ Z; —Val—AA$\overset{\downarrow}{-}$ Z; —Leu—AA$\overset{\downarrow}{-}$ Z; —Ile—AA$\overset{\downarrow}{-}$ Z
Pepsin	—Phe(or Tyr,Leu)$\overset{\downarrow}{-}$ Trp(or Phe,Tyr)—
Thermolysin	—AA$\overset{\downarrow}{-}$ Leu—; —AA$\overset{\downarrow}{-}$ Phe—; —AA$\overset{\downarrow}{-}$ Ile—; —AA$\overset{\downarrow}{-}$ Val—
Trypsin	—Arg$\overset{\downarrow}{-}$ Z; —Lys$\overset{\downarrow}{-}$ Z

[a] AA is any amino acid residue and Z is an amino acid residue, ester or amide. The cleavage sites (↓) are those preferred by the pure enzyme; crude preparations may have much broader specificities.

reaction mixture is produced, giving relatively small yields of BOC-aspartame. However, if two equivalents of the phenylalanine methyl ester are used, an insoluble addition complex forms in high yield at concentrations above 1 M. The loss of product from the liquid phase due to this precipitation greatly increases the overall yield of this process. Later, the BOC-aspartame may be released from this adduct by simply altering the pH. The stereospecificity of the thermolysin determines that only the L-isomer of phenylalanine methyl ester reacts but the addition product is formed equally well from both the D- and L-isomers. This fortuitous state of affairs allows the use of racemic phenylalanine methyl ester, the L-isomer being converted to the aspartame derivative and the D-isomer forming the insoluble complex shifting the equilibrium to product formation. D-Phenylalanine methyl ester released from the addition complex may be isomerised enzymically to reform the racemic mixture. The BOC-aspartame may be deprotected by a simple hydrogenation process to form aspartame:

BOC-L-aspartic acid + L-phenylalanine methyl ester

$\Updownarrow$ thermolysin

BOC-L-aspartame   [7.8]

$\downarrow$ D-phenylalanine methyl ester

BOC-L-aspartame : D-phenylalanine methyl ester adduct (precipitates)

Immobilised thermolysin cannot be used in this process as it has been found to co-precipitate with the insoluble adduct. However, this may be circumvented by its use within a liquid–liquid biphasic system.

The synthesis of aspartame is a very simple example of how proteases may be used in peptide synthesis. Most proteases show specificity in their cleavage sites (Table 7.2) and may be used to synthesise specific peptide linkages. Factors that favour peptide synthesis are correct choice of pH, the selection of protecting residues for amino and carboxyl groups that favour product precipitation and the use of liquid–liquid biphasic systems, all of which act by controlling the equilibrium of the reaction. An alternative strategy is kinetically controlled synthesis where the rate of peptide product synthesis (k_P) is high compared with the rate of peptide hydrolysis (k_H). This may be ensured by providing an amino acid or peptide which is a more powerful nucleophile than water in accepting a peptide unit from an enzyme–peptide intermediate. This kinetically controlled reaction may be represented as:

$$
\underset{\substack{\|\\\text{O}}}{\text{R}^1-\text{C}}-\underset{\substack{|\\\text{H}}}{\text{X}} + \text{enzyme} \;\rightleftharpoons\; [\underset{\substack{\|\\\text{O}}}{\text{R}^1-\text{C}}-\underset{\substack{|\\\text{H}}}{\text{X}}: \text{enzyme}] \longrightarrow \underset{\substack{\|\\\text{O}}}{\text{R}^1-\text{C}}-\text{enzyme} + \text{XH}
$$

$$
\text{H}_2\text{O}\!\!\downarrow k_H \qquad k_P\!\!\downarrow \quad \text{R}^2-\text{NH}_2
$$

$$
\underset{\substack{\|\\\text{O}}}{\text{R}^1-\text{COH}} \qquad\qquad \underset{\substack{\|\quad|\\\text{O}\;\;\text{H}}}{\text{R}^1-\text{C}-\text{N}-\text{R}^2}
$$

[7.9]

where X represents an alcohol, amine or other activating group, i.e. the reactant is an ester, amide (peptide) or activated carboxylic acid. The relative rate of peptide formation compared with hydrolysis depends on the ratio k_P/k_H, the ratio of the K_m values for water and amine, and the relative concentrations of the (unprotonated) amine and water (see equation (1.91), p. 31). Thus, where necessary, the reaction yield may be improved by lowering the water activity. It has been found that the yields may be increased by reducing the temperature to 4 °C, perhaps by a disproportionate effect on the K_m values. The amine and enzyme concentrations should be as high as possible for such kinetically controlled reactions and only those enzymes which utilise a covalently linked enzyme–peptide intermediate can be used (see Table 7.2). Also, the reaction is stopped well before equilibrium is reached as, under such thermodynamic control, the product peptide will be converted back though the enzyme intermediate to the carboxylic acid, a process made almost irreversible by its ionisation in solutions of pH above its pK_a, as shown in reaction scheme [7.10]

Peptides may be lengthened either by the addition of single amino acid residues or by the condensation of peptide fragments. The size of the peptide required as the final product may determine the type of synthesis used;

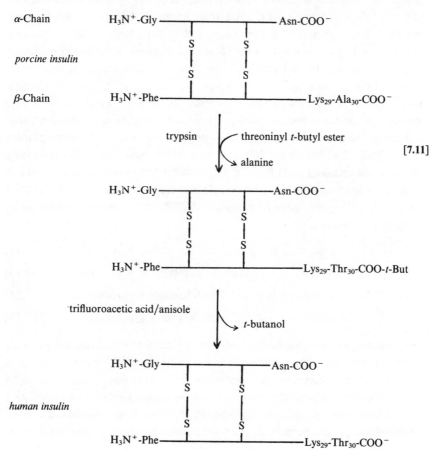

$$R^1-\overset{\overset{\displaystyle O}{\|}}{C}-X + \text{enzyme} \rightleftharpoons [R^1-\overset{\overset{\displaystyle O}{\|}}{C}-\overset{\overset{\displaystyle H}{|}}{X}: \text{enzyme}] \rightleftharpoons R^1-\overset{\overset{\displaystyle O}{\|}}{C}-\text{enzyme} + XH$$

$$H^{+'} + R^1-\overset{\overset{\displaystyle O}{\|}}{C}O^- \xleftarrow{\text{fast}} R^1-\overset{\overset{\displaystyle O}{\|}}{C}OH \qquad R^1C-\overset{\overset{\displaystyle O}{\|}}{N}-R^2 \qquad [7.10]$$

condensation of fragments may be performed in kinetically controlled processes, whereas stepwise elongation is best achieved using biphasic solid–liquid (i.e. precipitation) or liquid–liquid thermodynamically controlled processes.

In general, few proteases are required for such synthetic purposes. As they are quite costly, especially at the high activities necessary for kinetic control, they may be immobilised in order to enable their repeated use. Many examples of enzymic peptide synthesis may be cited. The conversion of porcine insulin to human insulin requires the replacement of the C-terminal

residue by a threonine. This can be achieved by a single
n step catalysed by trypsin using a carboxyl-protected threo-
ous solution with an organic co-solvent. The protective group
emoved later by mild hydrolysis and the product purified by
latography (see scheme [7.11])

Glycosidases used in synthetic reactions

This is a comparatively neglected topic, probably because polysaccharides
can be obtained readily from plant or microbial sources, because polysaccharide function is not so specifically related to its structure, and because the
theory describing the functional significance of oligosaccharides is still being
developed. As there is an abundance of knowledge about glycosidases and
their specificities, there is no fundamental reason why they should not be
used to synthesise oligosaccharides. There are few examples in the literature
of oligosaccharides being constructed using enzymes such as dextransucrase
(EC 2.4.1.5), levansucrase (EC 2.4.1.10), cyclomaltodextrin glucanotransferase (EC 2.4.1.19) and β-galactosidase (EC 3.2.1.23) to build up glucose,
fructose or galactose residues on suitable acceptors. Fructosyltransferases
(e.g. inulosucrase, EC 2.4.1.9) have been used to synthesise compounds such
as sucrose-6-acetate, xylosucrose (β-D-fructofuranosyl-(2,1)-α-D-xylopyra-
noside) and the low cariogenic sweetener 'neosugar', which consists of a
mixture of glucose, sucrose and β-2,1-linked fructans with terminal non-
reducing glucose residues (typical composition before chromatographic
refinement: 2% fructose, 26% glucose, 11% sucrose, 30% 1-kestose
(β-D-fructofuranosyl-(2,1)-β-D-fructofuranosyl-(2,1)-α-D-glucopyranose),
25% nystose (β-D-fructofuranosyl-(2,1)-β-D-fructofuranosyl-(2,1)-β-D-fruc-
tofuranosyl-(2,1)-α-D-glucopyranose) and 6% higher oligosaccharides, all by
weight):

$$\text{sucrose} + \text{glucose-6-acetate} \rightleftharpoons \text{sucrose-6-acetate} + \text{glucose} \qquad [7.12]$$

$$\text{sucrose} + \text{xylose} \rightleftharpoons \text{xylosucrose} + \text{glucose} \qquad [7.13]$$

$$\text{sucrose} + \text{sucrose} \rightleftharpoons \text{1-kestose} + \text{glucose} \qquad [7.14]$$

$$\text{1-kestose} + \text{sucrose} \rightleftharpoons \text{nystose} + \text{glucose} \qquad [7.15]$$

As shown in reaction scheme [7.16], dextransucrase can be used to produce
dextran (a polymer of α-1,6-linked glucose residues) from sucrose. Dextran is
used in gel chromatographic media, such as Sephadex, in the aqueous
two-phase systems already described, and as a blood plasma extender. For
any of these uses it is undesirable to have cells remaining in the dextran so the
dextransucrase, produced extracellularly by *Leuconostoc mesenteroides*, is

purified before use. The size of dextran molecules produced may be controlled by including small concentrations of sugars, such as maltose, which compete with sucrose as acceptors for glucose residues from the donor sucrose molecules:

$$\text{sucrose} + \text{dextran } (G_n) \rightleftharpoons \text{fructose} + \text{dextran } (G_{n+1}) \qquad [7.16]$$

Thermodynamically controlled synthesis of oligosaccharides, catalysed by glycosidases, is possible by use of high substrate concentrations and a 'molecular trap' to remove the products. For example, β-galactosidase may be used to synthesise *N*-acetyllactosamine (β-D-galactopyranosyl-(1,6)-2-acetamido-2-deoxy-D-glucose) using a carbon-celite column to remove it, as formed:

$$\text{galactose} + N\text{-acetylglucosamine} \underset{}{\overset{\text{β-galactosidase}}{\rightleftharpoons}} N\text{-acetyllactosamine} + H_2O$$

$$\downarrow \text{carbon-celite}$$

$$[7.17]$$

$$N\text{-acetyllactosamine}$$
$$(\text{trapped})$$

There seems to have been no systematic attempt to construct hetero-oligosaccharides in a stepwise movement, yet thermodynamically or kinetically controlled reactions could be used just as in peptide synthesis.

Interesterification of lipids

There is no opportunity or established need to build up polymers or oligomers using lipases or esterases, yet it is possible and commercially advantageous to use these enzymes as transferases in transesterifications (carboxyl group exchange between esters), acidolyses (carboxyl group exchange between esters and carboxylic acids) and alcoholyses (alcohol exchange between esters and alcohols). Certain triglycerides, cocoa butter being the outstanding example, have high value because of their physical properties and comparative rarity. There is a commercial pull, therefore, to find routes to the production of high-value triglycerides from more plentiful, cheaper raw materials. This may be done using lipases (e.g. *ex Rhizopus* or *ex Mucor miehei*) acting as transacylases, see [7.18].

This acidolysis reaction may be used for increasing the value of rapeseed oil by exchanging linoleic acid for linolenic acid residues and increasing the value of palm oil and sunflower oil by increasing their content of oleic acid residues. Cocoa butter is a relatively expensive fat, used in confectionery, because of its sharp melting point between room temperature and body temperature; chocolate literally melts in the mouth. This is due to the fairly

$$
\begin{array}{l}
\quad\ \ \overset{\displaystyle O}{\overset{\displaystyle \|}{H_2COCR^1}} \qquad\qquad\qquad \overset{\displaystyle O}{\overset{\displaystyle \|}{H_2COCR^4}} \\[4pt]
\overset{\displaystyle O}{\overset{\displaystyle \|}{R^2COCH}} \ + 2R^4\overset{\displaystyle O}{\overset{\displaystyle \|}{C}}OH \rightleftharpoons R^2\overset{\displaystyle O}{\overset{\displaystyle \|}{C}}OCH \ + R^1\overset{\displaystyle O}{\overset{\displaystyle \|}{C}}OH + R^3\overset{\displaystyle O}{\overset{\displaystyle \|}{C}}OH \qquad [7.18] \\[4pt]
\quad\ \ \overset{\displaystyle O}{\overset{\displaystyle \|}{H_2COCR^3}} \qquad\qquad\qquad \overset{\displaystyle O}{\overset{\displaystyle \|}{H_2COCR^4}}
\end{array}
$$

small variation in the structure of the constituent triglycerides; 80% have palmitic acid or stearic acid in the 1 and 3 positions, with oleic acid in the central 2 position. For the production of cocoa butter substitute from palm oil, a process which increases the value of the product three-fold, the acidolysis utilises stearic acid (i.e. R^4 is $C_{17}H_{35}$) in hexane containing just sufficient water to activate the lipase. Olive oil may be similarly improved by exchanging its 1,3-oleic acid residues for palmityl groups. The products may be recovered by recrystallisation from aqueous acetone. Such reactions may also be used for the resolution of racemic mixtures of carboxylic acids or alcohols, the lipase generally being specific for only one out of a pair of optical isomers.

The secret of success has been the selection of lipases with the correct specificity and the selection of reaction conditions that favour transacylation rather than hydrolysis. Because the hydrolytic activity of industrial lipases is 10–15 times the transacylation activity, it is advantageous to minimise the water content of the reaction system and use the aqueous–organic biphasic systems described earlier.

An example of lipase being used in an alcoholysis reaction is the biphasic production of isoamyl acetate, a natural aroma:

$$
CH_3CH_2OCOCH_3 + (CH_3)_2CHCH_2CH_2OH \overset{\text{lipase}}{\rightleftharpoons} (CH_3)_2CHCH_2CH_2OCOCH_3 +
$$
$$
CH_3CH_2OH \quad [7.19]
$$

ethyl acetate + isoamylalcohol $\rightleftharpoons$ isoamyl acetate + ethanol

Summary

(*a*) Reaction equilibria can be shifted by use of biphasic systems. This may be utilised by enzyme technologists so long as their enzymes are stable within such systems.

(*b*) Log P is the best measure of the polarity of a solution for use in enzymic reactions.

(*c*) Kinetic control may be preferred to thermodynamic control in some enzymic syntheses.

Bibliography

Carrea, G., Riva, S., Bovara, R. & Pasta, P. (1988). Enzymatic oxidoreduction of steroids in two-phase systems: effects of organic solvents on enzyme kinetics and evaluation of the performance of different reactors. *Enzyme and Microbial Technology*, **10**, 333–40.

Deetz, J. S. & Rozzell, J. D. (1988). Enzyme-catalysed reactions in non-aqueous media. *Trends in Biotechnology*, **6**, 15–19.

Klibanov, A. M. (1986). Enzymes that work in organic solvents. *CHEMTECH*, **16**, 354–9.

Laane, C. Boeren, S., Vos, K. & Veeger, C. (1987). Rules for optimization of biocatalysts in organic solvents. *Biotechnology and Bioengineering*, **30**, 81–7.

Lilly, M. D. (1982). Two-liquid-phase biocatalytic reactions. *Journal of Chemical Technology and Biotechnology*, **32**, 162–9.

Luisi, P. L. & Laane, C. (1986). Solubilization of enzymes in apolar solvents via reverse micelles. *Trends in Biotechnology*, **4**, 153–60.

Martinek, K. & Semenov, A. N. (1981). Enzymic synthesis in biphasic aqueous-organic systems. 2. Shift in ionic equilibria. *Biochimica et Biophysica Acta*, **658**, 90–101.

Martinek, K., Semenov, A. N. & Berezin, I. V. (1981). Enzymic synthesis in biphasic aqueous–organic systems. 1. Chemical equilibrium shift. *Biochimica et Biophysica Acta*, **658**, 76–89.

Morihara, K. (1987). Using proteases in peptide synthesis. *Trends in Biotechnology*, **5**, 164–70.

Nilsson, K. G. I. (1988). Enzymatic synthesis of oligosaccharides. *Trends in Biotechnology*, **6**, 256–64.

Rekker, R. F. & de Kort, H. M. (1979). The hydrophobic fragmental constant: an extension to a 1000 data point set. *European Journal of Medicinal Chemistry*, **14**, 479–88.

Visuri, K. & Klibanov, A. M. (1987). Enzymic production of high fructose corn syrup (HFCS) containing 55% fructose in aqueous ethanol. *Biotechnology and Bioengineering*, **30**, 917–20.

Zaks, A., Empie, M. & Gross, A. (1988). Potentially commerical enzymatic processes for the fine and speciality chemical industries. *Trends in Biotechnology*, **6**, 272–5.

Zaks, A. & Klibanov, A. M. (1984). Enzymic catalysis in organic media at 100 °C. *Science*, **224**, 1249–51.

Zaks, A. & Klibanov, A. M. (1985). Enzyme-catalyzed processes in organic solvents. *Proceedings of the National Academy of Sciences, USA*, **82**, 3192–6.

8   Future prospects for enzyme technology

Whither enzyme technology?

There are many directions in which enzyme technologists are currently applying their art and which are at the forefront of biotechnological research and development. Some of these have already been examined in some detail earlier (see Chapters 6 and 7). At present, relatively few enzymes are available on a large scale (i.e. > kg) and are suitable for industrial applications. These shortcomings are being addressed in a number of ways: (1) new enzymes are being sought in the natural environment and by strain selection (see Chapter 2); (2) established industrial enzymes are being used in as wide a variety of ways as can be conceived; (3) novel enzymes are being designed and produced by genetic engineering; (4) new organic catalysts are being designed and synthesised using the 'knowhow' established from enzymology; and (5) more complex enzyme systems are being utilised. Each of these areas has an extensive and rapidly expanding literature. Some advances possibly belong more properly to other areas of science. Thus, the development of genetically improved enzymes is generally undertaken by molecular biologists and the design and synthesis of novel enzyme-like catalysts is in the provenance of the organic chemists. Both groups of workers will, however, base their science on data provided by the enzyme technologist. Space requirements in this volume do not allow the full treatment of these related areas but will be discussed briefly here.

Use of 'unnatural' substrates

Many enzymes are not totally specific for their natural substrates. Some have been found to catalyse reactions quite different from those given as their normal reactions and reflected in their name and EC number. Sometimes it is necessary to place the enzyme in an unusual environment in order to display new activities. Thus, lipases act as transesterases in primarily non-aqueous environments (see Chapter 7). In other cases, changes in the environment are not necessary.

Glucose oxidase (see reaction scheme [**1.1**]) is specific for its reducing substrate (D-glucose) but fairly non-specific in its choice of oxidant, normally molecular oxygen. It has been established that benzoquinone is also an effective electron-accepting substrate:

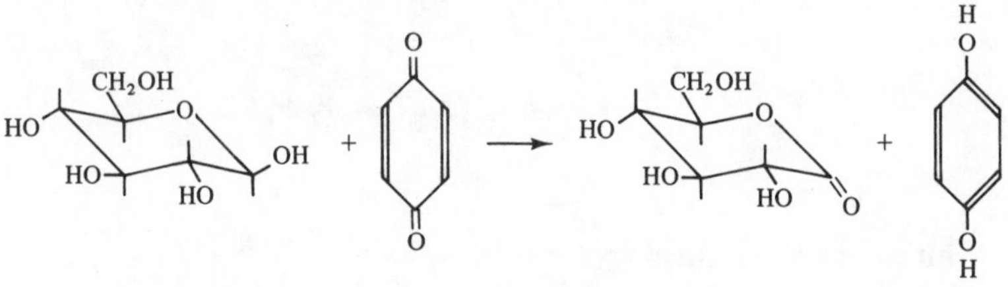

[8.1]

β-D-glucose + benzoquinone ⟶ D-glucono-1,5-lactone + hydroquinone

The product of this reaction, hydroquinone, is a valuable organic chemical, being used in the photographic industry and as an antioxidant. The reaction gives nearly 100% yields, with no possibility of the peroxide-induced inactivation which occurs using molecular oxygen as oxidant. Because of the ready solubility of benzoquinone and low solubility of molecular oxygen, the above reaction ([**8.1**]) can give productivity rates several times greater than the 'natural' reaction ([**1.1**]).

Acetylcholinesterase (EC 3.1.1.7) normally catalyses the hydrolysis of acetylcholine, the excitatory neurotransmitter, in the synaptic junctions of vertebrates.

$$(CH_3)_3\overset{+}{N}-CH_2-\underset{\displaystyle OCOCH_3}{\underset{|}{CH_2}} + H_2O \longrightarrow (CH_3)_3\overset{+}{N}-CH_2-\underset{\displaystyle OH}{\underset{|}{CH_2}} + CH_3COOH \qquad [8.2]$$

acetylcholine + water ⟶ choline + acetic acid

The acetylcholinesterase from the electric eel has been found additionally to catalyse the stereospecific hydrolysis of acetyl-D-carnitine but not acetyl-L-carnitine.

$$(CH_3)_3\overset{+}{N}-CH_2-\underset{\displaystyle OCOCH_3}{\underset{|}{CH}}-CH_2-COOH + H_2O \longrightarrow (CH_3)_3\overset{+}{N}-CH_2-\underset{\displaystyle OH}{\underset{|}{CH}}-CH_2-COOH + CH_3COOH \qquad [8.3]$$

acetyl-D-carnitine + water ⟶ D-carnitine + acetic acid

Although the reaction involving acetyl-D-carnitine has a second-order rate (specificity) constant four orders of magnitude smaller than that utilising acetylcholine, the productivities at high substrate concentrations are comparable, the 'natural' substrate, acetylcholine, causing pronounced substrate inhibition which is not apparent with acetyl-D-carnitine. The 'unnatural' reaction is useful as it may be used in the preparation of L-carnitine, one of the vitamins which has numerous therapeutic applications; the D-isomer being biologically inactive. The enzyme can, therefore, be used in a manner similar to the way in which aminoacylase is used to resolve racemic amino acids. Chemically synthesised racemic DL-carnitine may be acetylated, using acetyl chloride, to give acetyl-DL-carnitine. The acetylcholinesterase may then be used to produce a mixture of acetyl-L-carnitine and D-carnitine which may be simply resolved by ion-exchange chromatography. The acetyl-L-carnitine is as biologically active as L-carnitine and may be used directly.

Enzyme engineering

A most exciting development over the last few years is the application of *genetic engineering* techniques to enzyme technology. A full description of this burgeoning science is beyond the scope of this text but some suitable references are given at the end of this chapter. There are a number of properties which may be improved or altered by genetic engineering, including the yield and kinetics of the enzyme, the ease of downstream processing and various safety aspects. Enzymes from dangerous or unapproved microorganisms and from slow growing or limited plant or animal tissue may be cloned into safe high-production microorganisms. In the future, enzymes may be redesigned to fit more appropriately into industrial processes; for example, making glucose isomerase less susceptible to inhibition by the Ca^{2+} present in the starch saccharification processing stream.

The amount of enzyme produced by a microorganism may be increased by increasing the number of gene copies which code for it. This principle has been used to increase the activity of penicillin-G-amidase in *Escherichia coli*. The cellular DNA from a producing strain is selectively cleaved by the restriction endonuclease *Hind*III. This hydrolyses the DNA at relatively rare sites containing the 5′–AAGCTT–3′ base sequence to give identical 'staggered' ends.

$$
\begin{array}{ccc}
\begin{array}{l}
5' -A-A-G-C-T-T- 3' \\
\ \ \ | \ \ | \ \ | \ \ | \ \ | \ \ | \\
3' -T-T-C-G-A-A- 5'
\end{array}
&
\xrightarrow{\textit{Hind}\text{III}}
&
\begin{array}{l}
5' -A \qquad\qquad A-G-C-T-T- 3' \\
\ \ \ | \qquad\qquad\qquad\qquad\quad | \\
3' -T-T-C-G-A \qquad\qquad A- 5'
\end{array}
\end{array}
\quad +
\qquad [8.4]
$$

intact DNA   cleaved DNA

The total DNA is cleaved into about 10 000 fragments, only one of which contains the required genetic information. These fragments are individually cloned into a cosmid vector and thereby returned to *E. coli*. These colonies

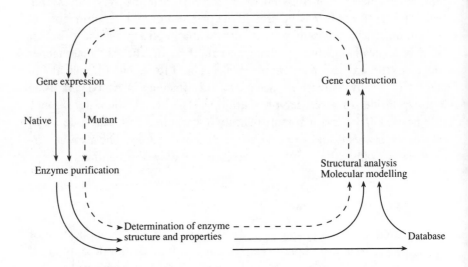

Figure 8.1. The protein engineering cycle. The process starts with the isolation and characterisation of the required enzyme. This information is analysed together with the database of known and putative structural effects of amino acid substitutions to produce a possible improved structure. This factitious enzyme is constructed by site-directed mutagenesis, isolated and characterised. The results, successful or unsuccessful, are added to the database, and the process repeated until the required result is obtained.

containing the active gene are identified by their inhibition of a 6-amino-penicillanic acid-sensitive organism. Such colonies are isolated and the penicillin-G-amidase gene transferred on to pBR322 plasmids and recloned back into *E. coli*. The engineered cells, aided by the plasmid amplification at around 50 copies per cell, produce penicillin-G-amidase constitutively and in considerably higher quantities than does the fully induced parental strain. Such increased yields are economically relevant not just for the increased volumetric productivity but also because of reduced downstream processing costs, the resulting crude enzyme being that much purer.

   Another extremely promising area of genetic engineering is *protein engineering*. New enzyme structures may be designed and produced in order to improve on existing enzymes or create new activities. An outline of the process of protein engineering is shown in Figure 8.1. Such factitious enzymes are produced by *site-directed mutagenesis* (Figure 8.2). Unfortunately from a practical point of view, much of the research effort in protein engineering has gone into studies concerning the structure and activity of enzymes chosen for their theoretical importance or ease of preparation rather than industrial relevance. This emphasis is likely to change in the future.

As indicated by the method used for site-directed mutagenesis (Figure 8.2), the preferred pathway for creating new enzymes is by the stepwise substitution of only one or two amino acid residues out of the total protein structure. Although a large database of sequence–structure correlations is available, and growing rapidly together with the necessary software, it is presently insufficient accurately to predict three-dimensional changes as a result of such substitutions. The main problem is assessing the long-range effects, including solvent interactions, on the new structure. As the many reported results will attest, the science is at a stage where it can explain the structural consequences of amino acid substitutions after they have been determined but cannot accurately predict them. Protein engineering, therefore, is presently rather a hit or miss process which may be used with only little realistic likelihood of immediate success. Apparently quite small sequence changes may give rise to large conformational alterations and even affect the rate-determining step in the enzymic catalysis. However it is reasonable to suppose that, given a sufficiently detailed database plus suitable software, the relative probability of success will increase over the coming years and the products of protein engineering will make a major impact on enzyme technology.

Much protein engineering has been directed at subtilisin (from *Bacillus amyloliquefaciens*), the principal enzyme in the detergent enzyme preparation, Alcalase. This has been aimed at the improvement of its activity in detergents by stabilising it at even higher temperatures, pH and oxidant strength. Most of the attempted improvements have concerned alterations to: (1) the P_1 cleft, which holds the amino acid on the carbonyl side of the targeted peptide bond; (2) the oxyanion hole (principally Asn155), which stabilises the tetrahedral intermediate; (3) the neighbourhood of the catalytic histidyl residue (His64), which has a general base role; and (4) the methionine residue (Met222) which causes subtilisin's lability to oxidation. It has been found that the effect of a substitution in the P_1 cleft on the relative specific activity between substrates may be fairly accurately predicted even though predictions of the absolute effects of such changes are less successful. Many substitutions, particularly for the glycine residue at the bottom of the P_1 cleft (Gly166), have been found to increase the specificity of the enzyme for particular peptide links whilst reducing it for others. These effects are achieved mainly by corresponding changes in the K_m rather than the V_{max}. Increases in relative specificity may be useful for some applications. They should not be thought of as the usual result of engineering enzymes, however, as native subtilisin is unusual in being fairly non-specific in its actions,

(a)

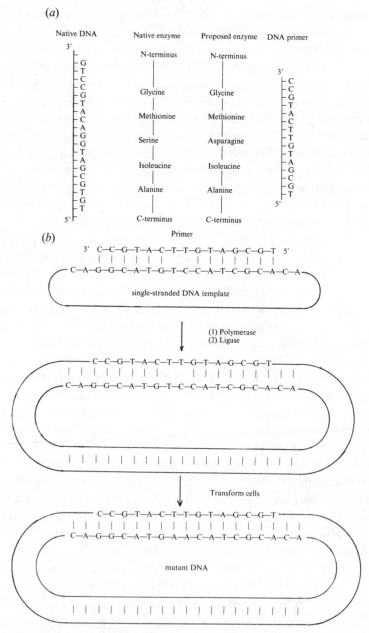

(b)

Figure 8.2. An outline of the process of site-directed mutagenesis, using a hypothetical example. (a) The primary structure of the enzyme is derived from the DNA sequence. A putative enzyme primary structure is proposed with an asparagine residue replacing the serine present in the native enzyme. A short piece of DNA (the primer), complementary to a section of the gene apart from the base mismatch, is synthesised. (b) The

possessing a large hydrophobic binding site which may be made more specific relatively easily (e.g. by reducing its size). The inactivation of subtilisin in bleaching solutions coincides with the conversion of Met222 to its sulphoxide, the consequential increase in volume occluding the oxyanion hole. Substitution of this methionine by serine or alanine produces mutants which are relatively stable, although possessing somewhat reduced activity.

An example of the unpredictable nature of protein engineering is given by trypsin, which has an active site closely related to that of subtilisin. Substitution of the negatively charged aspartic acid residue at the bottom of its P_1 cleft (Asp189), which is used for binding the basic side-chains of lysine or arginine, by positively charged lysine gives the predictable result of abolishing the activity against its normal substrates but unpredictably also gives no activity against substrates where these basic residues are replaced by aspartic acid or glutamic acid.

Considerable effort has been spent on engineering more thermophilic enzymes. It has been found that thermophilic enzymes are generally only 20–30 kJ more stable than their mesophilic counterparts. This may be achieved by the addition of just a few extra hydrogen bonds, an internal salt link or extra internal hydrophobic residues, giving a slightly more hydrophobic core. All of these changes are small enough to be achieved by protein engineering. To ensure a more predictable outcome, the secondary structure of the enzyme must be conserved and this generally restricts changes in the exterior surface of the enzyme. Suitable for exterior substitutions for increasing thermostability have been found to be aspartate → glutamate, lysine → glutamine, valine → threonine, serine → asparagine, isoleucine → threonine, asparagine → aspartate and lysine → arginine. Such substitutions have a fair probability of success. Where allowable, small increases in the interior hydrophobicity for example by substituting interior glycine or serine residues by alanine may also increase the thermostability. It should be recognised that making an enzyme more thermostable reduces its overall flexibility and, hence, it is probable that the factitious enzyme produced will have reduced catalytic efficiency.

Figure 8.2 (*cont.*)

oligonucleotide primer is annealed to a single-stranded copy of the gene and is extended with enzymes and nucleotide triphosphates to give a double-stranded gene. On reproduction, the gene gives rise to both mutant and wild-type clones. The mutant DNA may be identified by hybridisation with radioactively labelled oligonucleotides of complementary structure.

Artificial enzymes

A number of possibilities now exist for the construction of artificial enzymes. These are generally synthetic polymers or oligomers with enzyme-like activities, often called *synzymes*. They must possess two structural entities, a substrate-binding site and a catalytically effective site. It has been found that producing the facility for substrate binding is relatively straightforward but catalytic sites are somewhat more difficult. Both sites may be designed separately but it appears that, if the synzyme has a binding site for the reaction transition state, this often achieves both functions. Synzymes generally obey the saturation Michaelis–Menten kinetics as outlined in Chapter 1. For a one-substrate reaction the reaction sequence is given by

$$\text{synzyme} + S \rightleftharpoons (\text{synzyme–S complex}) \rightarrow \text{synzyme} + P \qquad [8.5]$$

Some synzymes are simply derivatised proteins, although covalently immobilised enzymes are not considered here. An example is the derivatisation of myoglobin, the oxygen carrier in muscle, by attaching $(Ru(NH_3)_5)^{3+}$ to three surface histidine residues. This converts it from an oxygen carrier to an oxidase, oxidising ascorbic acid whilst reducing molecular oxygen. The synzyme is almost as effective as natural ascorbate oxidases.

It is impossible to design protein synzymes from scratch with any probability of success, as their conformations are not presently predictable from their primary structure. Such proteins will also show the drawbacks of natural enzymes, being sensitive to denaturation, oxidation and hydrolysis. For example, polylysine binds anionic dyes but only 10% as strongly as the natural binding protein, serum albumin, in spite of the many charges and apolar side-chains. Polyglutamic acid, however, shows synzymic properties. It acts as an esterase in much the same fashion as the acid proteases, showing a bell-shaped pH–activity relationship, with optimum activity at about pH 5.3, and Michaelis–Menten kinetics with a K_m of 2 mM and V_{max} of 10^{-4} to 10^{-5} s^{-1} for the hydrolysis of 4-nitrophenyl acetate.

Cyclodextrins (Schardinger dextrins) are naturally occurring toroidal molecules consisting of six, seven, eight, nine or ten α–1, 4-linked D-glucose units joined head-to-tail in a ring (α-, β-, γ-, δ- and ϵ-cyclodextrins, respectively: they may be synthesised from starch by the cyclomaltodextrin glucanotransferase (EC 2.4.1.19) from *Bacillus macerans*). They differ in the diameter of their cavities (about 0.5–1 nm) but all are about 0.7 nm deep. These form hydrophobic pockets due to the glycosidic oxygens and CH groups facing inwards. All the C–6 hydroxyl groups project to one end and all the C–2 and C–3 hydroxyl groups to the other. Their overall characteristic is hydrophilic, being water soluble, but the presence of their hydrophobic pocket enables them to bind hydrophobic molecules of the appropriate size. Synzymic cyclodextrins are usually derivatised in order to introduce catalytically relevant groups. Many such derivatives have been examined. For example, a C–6 hydroxyl group of β-cyclodextrin was covalently derivatised by an activated pyridoxal coenzyme. The resulting synzyme not only acted as a transaminase (see reaction scheme [1.2]) but also showed stereoselectivity for the L-amino acids. It was not as active as natural transaminases, however.

Polyethyleneimine is formed by polymerising ethyleneimine to give a highly branched hydrophilic three-dimensional matrix. About 25% of the resultant amines are primary, 50% secondary and 25% tertiary:

$$
4n \quad
\begin{array}{c}
CH_2 \\
| \quad \diagdown \\
| \quad \diagup \text{NH} \\
CH_2
\end{array}
\qquad
\left[
\begin{array}{l}
-CH_2-N-CH_2-CH_2-NH-CH_2- \\
\qquad\qquad | \\
\quad CH_2-CH_2-NH-CH_2-CH_2-NH_2
\end{array}
\right]_n
\qquad [8.6]
$$

ethyleneimine   polyethyleneimine

The primary amines may be alkylated to form a number of derivatives. If 40% of them are alkylated with 1-iodododecane to give hydrophobic binding sites and the remainder alkylated with 4(5)-chloromethylimidazole to give general acid–base catalytic sites, the resultant synzyme has 27% of the activity of α-chymotrypsin against 4-nitrophenyl esters. As might be expected from its apparently random structure, it has very low esterase specificity. Other synzymes may be created in a similar manner.

Antibodies to transition state analogues of the required reaction may act as synzymes. For example, phosphonate esters of general formula $(R\text{-}PO_2\text{-}OR')^-$ are stable analogues of the transition state occurring in carboxylic ester hydrolysis. Monoclonal antibodies raised to immunising protein conjugates covalently attached to these phosphonate esters act as esterases. The specificities of these catalytic antibodies (also called *abzymes*) depends on the structure of the side-chains (i.e. R and R' in $(R\text{-}PO_2-OR')^-$) of the antigens. The K_m values may be quite low, often in the micromolar region, whereas the V_{max} values are low (below $1\ s^{-1}$), although still 1000-fold higher than hydrolysis by background hydroxyl ions. A similar strategy may be used to produce synzymes by molecular 'imprinting' of polymers, using the presence of transition state analogues to shape polymerising resins or inactive non-enzymic protein during heat denaturation.

Coenzyme-regenerating systems

Many oxidoreductases and all ligases utilise coenzymes (e.g. NAD^+, $NADP^+$, NADH, NADPH, ATP), which must be regenerated as each product molecule is formed. Although these represent many of the most useful biological catalysts, their application is presently severely limited by the high cost of the coenzymes and difficulties with their regeneration. These two problems may both be overcome at the same time if the coenzyme is immobilised, together with the enzyme, and regenerated *in situ*.

A simple way of immobilising/regenerating coenzymes would be to use whole-cell systems and these are, of course, in widespread use. However as outlined earlier, these are of generally lower efficiency and flexibility than immobilised-enzyme systems. Membrane reactors (see Chapter 5) may be used to immobilise the coenzymes but the pore size must be smaller than the coenzyme diameter, which is extremely restrictive. Coenzymes usually must be derivatised for adequate immobilisation and regeneration. When successfully applied, this process activates the coenzymes for attachment to the immobilisation support but does not interfere with its biological function. The most widely applied synthetic routes involve the alkylation of the exocyclic N^6-amino nitrogen of the adenine moiety present in the coenzymes NAD^+, $NADP^+$, NADH, NADPH, ATP and coenzyme A.

In some applications, such as those using membrane reactors it is only necessary that the coenzyme has sufficient size to be retained within the system. High molecular weight water-soluble derivatives are most useful as they cause less diffusional resistance than insoluble coenzyme matrices. Dextrans, polyethyleneimine and polyethylene glycols are widely used. Relatively low levels of coenzyme attachment are generally sought in order to allow greater freedom of movement and avoid possible inhibitory effects. The kinetic properties of the derived coenzymes vary, depending upon the system, but generally the Michaelis constants are higher and the maximum velocities are lower than with the native coenzymes. Coenzymes immobilised to insoluble supports presently have somewhat less favourable kinetics even when co-immobilised close to the active site of their utilising enzymes. This situation is expected to improve as more information on the protein conformation surrounding the enzymes' active sites becomes available and immobilisation methods become more sophisticated. However, the cost of such derivatives is always likely to remain high and they will only be economically viable for the production of very high value products.

There are several systems available for the regeneration of the derivatised coenzymes by chemical, electrochemical or enzymic means. Enzymic regeneration is advantageous because of its high specificity but electrochemical procedures for regenerating the oxidoreductase dinucleotides are proving competitive. To be useful in regenerating coenzymes, enzymic processes must utilise cheap substrates and readily available enzymes and give non-interfering and easily separated products. Formate dehydrogenase and acetate kinase present useful examples of their use, although the presently available commercial enzyme preparations are of low activity:

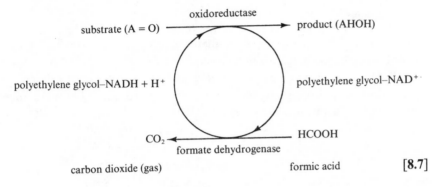

[8.7]

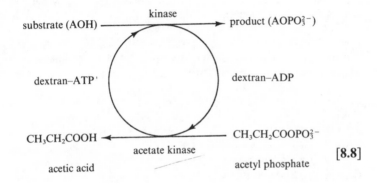

[8.8]

Conclusions

Enzyme technology is presently going through a phase of maturation and evolution. The maturation is shown by the development of the theory concerning how enzymes function and how this is related to their primary structure through the formation and configuration of their three-dimensional structure. The evolution is shown by the ever-broadening range of enzymic applications.

There still remains much room for the development of useful processes and materials based on this hard-won understanding. Enzymes will clearly be more widely used in the future and this will be reflected in the number of enzymes available on an industrial (and research) scale, the variety of reactions catalysed and the range of environmental conditions under which they will operate. Established enzymes will be put to new uses and novel enzymes, discovered within their biological niches or produced by design using enzyme engineering, will be used to catalyse hitherto unexploited reactions. This is just the start of the enzyme technology era.

Summary

(*a*) New enzymic processes are being developed and new uses are being devised for currently available industrial enzymes.

(*b*) New enzymes are being produced by classical means and 'by design'.

(*c*) Protein engineering offers great promise for the future but needs the development of a stronger theoretical base.

Bibliography

Cambou, B. & Klibanov, A. M. (1984). Unusual catalytic properties of usual enzymes. In *Enzyme engineering*, vol. 5, ed. A. I. Laskin, G. T. Tsao & L. B. Wingard Jr, pp. 219–23. New York: New York Academy of Sciences.

Lerner, R. A. & Tramontano, A., (1987). Antibodies as enzymes. *Trends in Biochemical Sciences*, **12**, 427–30.

Lowe, C. R. (1981). Immobilised coenzymes. In *Topics in enzyme and fermentation biotechnology*, vol. 5, ed. A. Wiseman, pp. 13–146. Chichester: Ellis Horwood Ltd.

Mänson, M.-A. & Mosbach, K. (1987). Immobilised active coenzymes. *Methods in Enzymology*, **136**, 3–9.

Mutter, M. (1988). Nature's rules and chemist's tools: a way for creating novel proteins. *Trends in Biochemical Sciences*, **13**, 260–5.

Peberdy, J. F. (1987). Genetic engineering in relation to enzymes. In *Biotechnology*, vol. 7a *Enzyme technology*, ed. J. F. Kennedy, pp. 325–44, Weinheim: VCH Verlagsgesellschaft mbH.

Pike, V. W. (1987). Synthetic enzymes. In *Biotechnology*, vol. 7a *Enzyme technology*, ed. J. F. Kennedy, pp. 465–85, Weinheim: VCH Verlagsgesellschaft mbH.

Pühler, A. & Heumann, W. (1981). Genetic engineering. In *Biotechnology*, vol. 1, ed. H.-J. Rehm & G. Reed, pp. 331–54, Weinheim: VCH Verlagsgesellschaft mbH.

Querol, E. & Parrila, A. (1987). Tentative rules for increasing the thermostability of enzymes by protein engineering. *Enzyme and Microbial Technology*, **9**, 238–244.

Shaw, W. V. (1987). Protein engineering: the design, synthesis and characterisation of factitious proteins. *Biochemical Journal*, **246**, 1–17.

Wells, J. A. & Estell, D. A. (1988). Subtilisin – an enzyme designed to be engineered. *Trends in Biochemical Sciences*, **13**, 291–7.

Winter, G. & Fersht, A. R. (1984). Engineering enzymes. *Trends in Biotechnology*, **2**, 115–19.

Wiseman, A. & Dalton, H., (1987). Enzymes versus enzyme-mimetic systems for biotechnological applications. *Trends in Biotechnology*, **5**, 241–4.

Index